AF842903

Contractor Safety Management

Contractors bear the brunt of workplace illnesses, injuries, and fatalities (~80%) of the estimated 3 million fatalities and 395 million non-fatal injuries annually. This book provides a series of proven practices, processes, and tools designed to protect the contracting workforce. In today's competitive landscape, organizations leverage contractors for flexibility and specialized expertise, yet this outsourcing introduces safety challenges that transcend traditional workplace boundaries. Failed contractor safety management brings severe consequences, preventable injuries, operational disruptions, regulatory penalties, reputational damage, and financial losses. Drawing from two decades of cross-industry experience, this book bridges the gap between contractual safety requirements and practical implementation, transforming transactional contractor relationships into strategic partnerships built on shared safety value-creating environments where safety excellence drives rather than competes with business performance.

Built on a methodology that has slashed incident rates by 70% in high-risk industries, *Contractor Safety Management: A Framework for Delivering Excellence in HSE Performance* delivers immediate actionable tools for HSE professionals and leaders, comprehensive prequalification templates, seamless onboarding protocols, barrier-breaking communication frameworks, and predictive reporting systems. Real-world case studies across construction, manufacturing, energy, and infrastructure sectors provide adaptable solutions for organizations of any size or safety maturity.

This comprehensive safety management guide is written for executives seeking a competitive advantage, operations managers aiming to boost productivity and reduce workplace incidents, procurement specialists requiring balanced selection criteria, and safety practitioners needing technical frameworks for contractor safety performance improvements. The book serves organizations spanning from single-site operations to global networks, helping transform traditional compliance-focused approaches into strategic safety partnerships that deliver measurable business value.

Contractor Safety Management

A Framework for Delivering Excellence in HSE Performance

Chitram Lutchman, Kevan Lutchman, and
Ramakrishna Akula

CRC Press
Taylor & Francis Group
Boca Raton London New York

CRC Press is an imprint of the
Taylor & Francis Group, an **informa** business

First edition published 2027
by CRC Press
2385 NW Executive Center Drive, Suite 320, Boca Raton FL 33431

and by CRC Press
4 Park Square, Milton Park, Abingdon, Oxon, OX14 4RN

CRC Press is an imprint of Taylor & Francis Group, LLC

© 2027 Chitram Lutchman, Kevan Lutchman, and Ramakrishna Akula

Reasonable efforts have been made to publish reliable data and information, but the author and publisher cannot assume responsibility for the validity of all materials or the consequences of their use. The authors and publishers have attempted to trace the copyright holders of all material reproduced in this publication and apologize to copyright holders if permission to publish in this form has not been obtained. If any copyright material has not been acknowledged please write and let us know so we may rectify in any future reprint.

Except as permitted under U.S. Copyright Law, no part of this book may be reprinted, reproduced, transmitted, or utilized in any form by any electronic, mechanical, or other means, now known or hereafter invented, including photocopying, microfilming, and recording, or in any information storage or retrieval system, without written permission from the publishers.

For permission to photocopy or use material electronically from this work, access www.copyright.com or contact the Copyright Clearance Center, Inc. (CCC), 222 Rosewood Drive, Danvers, MA 01923, 978-750-8400. For works that are not available on CCC please contact mpkbookspermissions@tandf.co.uk

For Product Safety Concerns and Information please contact our EU representative GPSR@taylorandfrancis.com. Taylor & Francis Verlag GmbH, Kaufingerstraße 24, 80331 München, Germany.

Trademark notice: Product or corporate names may be trademarks or registered trademarks and are used only for identification and explanation without intent to infringe.

ISBN: 978-1-041-21640-7 (hbk)
ISBN: 978-1-041-21677-3 (pbk)
ISBN: 978-1-003-72846-7 (ebk)

DOI: 10.1201/9781003728467

Typeset in ITC Stone Sans Std
by KnowledgeWorks Global Ltd.

This book is dedicated to every worker who has ever been injured or killed in the workplace, as well as their families. To the families of our lost colleagues and friends, we offer our most sincere condolences.

This book is also dedicated to the many HSE practitioners and leaders across the globe who spend countless hours seeking new ways for improving workplace safety and without whose commitment proven practices would not be available for generating meaningful safety performance improvements.

CONTENTS

PREFACE

This book, designed for contractors and organizations of all sizes, intends to provide a pathway to improving contractor safety performance broadly across various industries and workplaces. The content consolidates tried and proven workplace practices which have delivered performance improvements in Contractor Safety Management (CSM). We, like many Health, Safety and Environment (HSE) practitioners across the globe, share a common commitment to perpetual learning. Our hope is that the content of this book will strengthen the foundational knowledge of its readers to continue our advancements in contractor safety performance.

Contractors are the disproportionate recipients of workplace harm—historically experiencing approximately three out of four safety events. This reality is compounded by the general agreement that over 90% of workplace events are predictable and avoidable. We are very confident that a structured, organized, and disciplined approach to CSM will reduce the current global toll of approximately 3 million worker fatalities each year.

Protecting the contractor can be extremely difficult. They are expected to perform dirty, dangerous, and difficult work—and are expected to complete this work cheaper, better, and faster. Combine these factors with contractor inexperience due to transient roles, and the risk of errors and hazard exposures is undoubtedly heightened. An understanding of integrated management systems such as HSE, Process Safety Management (PSM), and Operational Excellence Management Systems (OEMS) becomes foundational to achieving safe outcomes.

We have combined our lenses of experience in HSE management, operations and project leadership, occupational health, organizational systems development, adult education, and technical writing for users. In so doing, we hope we have created a practical and streamlined guide for the reader seeking to face the many challenges associated with stewarding and managing contractor safety.

Together, we can reinforce a global culture of bolstering operational reliability while prioritizing human life.

We hope you find this to be a valuable resource.

AUTHORS

Chitram Lutchman is the Principal at OEMS Erudite Corporation. Dr. Lutchman is a hands-on energy professional with more than 35 years of experience in leading operations, business planning, EH&S, and integrated management systems (HSE, PSM, and OEMS). Chit is well experienced in project and operations risk management and is a published author in the areas of project management, EH&S, PSM, and OEMS. Dr. Lutchman is a leading expert on Contractor Safety Management (CSM) and has presented at conferences and led workshops on CSM internationally. Building on work experience in leading energy producers in North America, Chit seeks to bring business intelligence, operations discipline, risk management, and EH&S under a single umbrella to produce exceptional and sustainable organizational discipline and excellence. He is particularly passionate about improving CSM because of his belief that all workers are equally valued and no one should be hurt or fatally injured in the workplace and that workplace injuries and fatalities are highly skewed toward contractors. Over the past 20 years, working in CSM, he has introduced several practical applications (detailed in this book) to improve contractor safety performance.

As principal of OEMS Erudite Corporation, Chit seeks to provide quality and sustainable solutions to clients in developing and implementing fit-for-purpose IMS, leveraging expertise in the areas of leadership, risk management, EH&S, PSM, OEMS, and CSM.

Chit has provided solutions for improving operations discipline and managing operations risk and EH&S to ENOC, BAPCO, Suncor Energy, Husky Energy, AltaGas, Centrica Energy, MEG Energy, ARC Resources, Plains All American, HollyFrontier Corporation, and several others.

Dr. Lutchman is a Certified Safety Professional/Canadian Registered Safety Professional (CSP/CRSP). Today, Dr. Lutchman is regarded as a global and leading expert in the field of Contractor Safety Management in Canada, the United States, the Middle East, Asia, and the Caribbean.

Kevan Lutchman is a highly trained occupational and environmental medicine physician. He trained and subsequently instructed at the Harvard Occupational Medicine Residency (OEMR) program. In his current roles, Dr. Lutchman practices clinical medicine, conducts academic research, and oversees the training and mentorship of resident and postdoctoral research fellow physicians. His clinical expertise spans pre-placement medical evaluations, management of workplace injuries, fitness-for-duty determinations, and a range of specialized occupational health consultations. Dr. Lutchman also maintains double board certification, an active teaching role, and currently manages a Board of Directors position.

Dr. Lutchman's work reflects broad experience across diverse industries, including healthcare, construction, transportation, biopharmaceutical manufacturing, first responders, and waste management. His commitment to education has been recognized by the Harvard School of Public Health with teaching awards in 2024 and 2025, underscoring his dedication to developing future leaders in occupational and environmental medicine.

Beyond his clinical and teaching responsibilities, Dr. Lutchman has led several impactful initiatives. He designed and implemented a mental health program for a major healthcare organization, aimed at assessing and improving worker stress and overall wellbeing. He served as the primary subject-matter expert for a multinational company in developing a sleep physiology whitepaper and presenting recommendations on safe shift-work practices to reduce injuries and long-term health risks.

Dr. Lutchman also created a comprehensive evaluation framework for a mobile healthcare organization serving migrant workforces, including a logic model, survey data analysis, and a roadmap for program growth. He has been consulted in the development of health and safety initiatives for several organizations regarding the identification and management of workplace respiratory disease.

Ramakrishna Akula is a Chartered Fellow of the Occupational Safety and Health Practitioner (UK) and Certified Safety Professional (USA). Dr. Akula, also known as Dr. Rama, is a seasoned expert in the oil and gas operations and safety domain, boasting 25 years of experience. Throughout his illustrious career, he has worked with several prestigious international organizations, including ADNOC, ENOC, HPCL, and Aramco. Dr. Rama has extensive hands-on experience in managing contractor safety management, overseeing the entire lifecycle of contracts from scope development to completion. He is skilled in developing contractor scopes, evaluating contractor performance, and creating and managing key performance indicators (KPIs). Additionally, Dr. Rama has conducted numerous contractor performance evaluations, camp audits, and implemented contractor wellbeing programs, ensuring that contractors meet safety expectations.

Dr. Rama is a distinguished professional, holding esteemed certifications and fellowships. He is a Chartered Fellow of the Occupational Safety and Health Practitioner, a Fellow Member of the International Institute of Risk and Safety Management (UK), and a Certified Safety Professional. These accolades demonstrate his exceptional expertise and commitment to the field of occupational safety and health.

With his extensive experience in large integrated oil and gas companies, Dr. Rama has developed a deep understanding of contractor management processes and documentation. He has hands-on experience in managing contracts end-to-end, ensuring that all aspects of contractor management are meticulously executed. His expertise and dedication have made him a trusted authority in the field of operations and safety, and his contributions have significantly enhanced the safety and efficiency of contractor management processes in industry.

INTRODUCTION

"Buckle up."

We all know it refers to fastening one's seatbelt. Conversationally, it has been carried through history as a command to prepare oneself. Somehow, these two simple words are powerful enough to save lives. According to the National Highway Traffic Safety Administration (NHTSA), "from 1975 through 2017, seat belts have saved an estimated 374,276 lives" in the United States alone.[1] The World Health Organization and other international entities estimate that seat belts have saved millions of lives globally since their advent.[2] Driving is hazardous, but it is a necessary and regular endeavor for most of us. Consequently, we continually strive as a species to minimize the inherent life-threatening risks of modern transportation. Something as apparently trivial as fastening a thin strap across our torsos is so impactful that the act of doing so is directly translatable to dozens of languages across the globe. Effective safety appeals to just about everyone.

So why is it that an estimated 2.9 million workers die, and 395 million workers suffer work-related injuries or illnesses every year? These figures dwarf the Centers for Disease Control's 2023 global estimate of 1.19 million annual roadway fatalities.[3] These figures rival the impact of pandemics, which grip the attention of humanity and demand perpetual vigilance. Yet, this crisis receives so little headline consideration that one might assume we are simply okay with it. The global burden of workplace harm continues to smolder—underreported and underdiscussed.

Why is this so overlooked? Perhaps the daily toll of 10 to 12 workplace fatalities in the United States alone fails to reach the public alarm threshold. Perhaps the workplace death toll of developing nations—shrouded in inaccuracy—seems a little too geographically distant to matter. For many leaders, including ourselves, this silence is troubling. Despite our best efforts to improve workplace safety, injuries and fatalities are far too commonplace, undermining our core goal: *ensuring the safety of every worker.*

Organizations invest heavily in health, safety, and environment (HSE) management systems, employee training, personal protective equipment (PPE), emergency response plans, and employee wellbeing. Still, as industries demand more from workers in fast-changing environments, fatalities continue to occur from accidents and illness alike.

Safety slogans and buzzwords such as those below overflow the workplace:

- If we can't do it safely, we won't do it at all
- Zero Harm
- Target Zero
- Beyond Zero

While mantras and mottos can guide cultural direction, they don't always translate into real-world results. The truth is that context is more impactful than any set of words could ever be. Who was the first person to ever tell you to "buckle up"? For most people today, it was a parent. Those words—an instruction—were rooted in genuine care. Those words were spoken in the context of consistent protective parental actions and a strong personal bond. And we listened because we were reminded until it became reflexive, and because we watched our parents buckle up too. The NHTSA estimated that in the United States, 2024 saw 91.2% of adult front-seat passengers using seatbelts at any moment, a figure consistent with the approximately 90% usage rate since 2016.[4] Clearly, "buckle up" clicked.

In the workplace, we tend to care most when tragedy is personal. When our loved ones are killed or seriously injured at work, grief-based responses emerge only to be overtaken by competing priorities over time. Post-incident investigations can uncover valuable lessons—if they're done thoughtfully and with true intent. Sadly, legal considerations and internal policies limit information sharing. In many cases, affected families may receive financial compensation tethered to non-disclosure agreements, which further obscure accidents' root causes. Regulatory bodies may produce reports months or even years later, by which time the memory of the event has faded. By then, the worker has become a statistic whose suffering is remembered only by family and close friends.

Many leaders *do* genuinely strive to create safer workplaces. However, the scope and severity of occupational hazards constantly change and are never fully understood. Companies develop policies, procedures, and safety protocols and mandate PPE. Yet, workers still suffer.

Why? Maybe we prioritize efficiency over risk. In our drive to excel, maybe we look for shortcuts, bypassing procedures and foregoing PPE. Maybe we fail to adequately educate workers on the personal consequences of injury. Maybe we inadvertently reward risky behavior by praising production.

To you, the reader: we implore you to think of your family members (employee, contractor) who are out working daily. If you knew your loved ones were exposed to hazards regularly, would you advocate for their protection? We believe so.

Dr. Chit Lutchman: As a survivor of a workplace incident, I now remind my family constantly: "be safe" is not a cliché. It is a call to action, just like "buckle up."

To us, "be safe" means:

- Are you aware of today's assigned work and hazards?
- Are you trained and competent for the task?
- Do you have—and are you following—a procedure?
- Do you have the correct PPE—and are you using it?
- Are you alert and aware of your surroundings and related hazards and risks?
- Are you staying out of harm's way?
- Do you know what to do if something goes wrong?
- Are you a competent team member and are others competent too?

We say "be safe" to our loved ones instinctively. But when those words are spoken with intention and awareness, they become powerful reminders of what truly matters.

Thanks to the tireless efforts of regulators, unions, leaders, and frontline health and safety professionals, the workplace has become safer over time. We *can* improve; we've shown it. Workers are now more empowered to refuse unsafe work, and unsafe behaviors are routinely corrected. But challenges persist, especially for the oft overlooked contract workforce.

So why write this book?

Because we believe no one should be harmed at work, and everyone deserves to go home safely. Because we believe organizations can do more—especially for contractors—by adopting consistent and effective processes to enhance both safety and performance.

In this book, we share a series of proven, customizable practices, processes, and tools designed to support safe, reliable, and efficient contractor safety performance. We hope that these insights will meaningfully contribute to a global workplace culture where contractor safety management is not just a spoken priority, but a deeply held value shared by all.

Notes

1 National Highway Traffic Safety Administration. 2025. Seat Belt Safety: Buckle Up America. U.S. Department of Transportation. Accessed September 1, 2025. https://www.nhtsa.gov/vehicle-safety/seat-belts#:~:text=Make%20it%20a%20habit%20to%20always%20buckle,belts%20have%20saved%20an%20estimated%20374%2C276%20lives

2 World Health Organization. 2023. Road Traffic Injuries. Accessed September 1, 2025. https://www.who.int/news-room/fact-sheets/detail/road-traffic-injuries

3 World Health Organization. 2023. Global Status Report on Road Safety 2023. Geneva: World Health Organization. Accessed September 1, 2025. https://www.who.int/teams/social-determinants-of-health/safety-and-mobility/global-status-report-on-road-safety-2023

4 National Highway Traffic Safety Administration. 2025. *Seat Belt Safety: Buckle Up America*. U.S. Department of Transportation. Accessed September 1, 2025. https://www.nhtsa.gov/vehicle-safety/seat-belts

1 THE CASE FOR CHANGE – A GLOBAL PERSPECTIVE

Compelling Reasons for Improving Contractor Safety Management

Managing contractor health and safety is inherently challenging, especially as most organizations pursue the paradoxical goal of performing work *cheaper, better, and faster*. This workplace expectation has failed repeatedly, yet it remains embedded in our workplace culture. Nevertheless, significant progress has been made in contractor safety management (CSM). Many organizations are transitioning from a divided "us and them" mentality to shared accountability of safety performance in executing assigned work.

Today, contractors perform 75–80% of the work for most organizations in high-risk industries such as energy, construction, and manufacturing. To meaningfully improve workplace health, safety and environmental (HSE) performance, it's obvious that organizations must prioritize CSM. This requires two critical elements:

- Mindset changes among organizational leaders.
- A structured, sustainable approach to contractor HSE performance management.

Both elements require ongoing commitment to be effective.

According to the International Labour Organization (ILO, 2023), over 2.9 million workers died on the job in 2019, with 395 million more suffering nonfatal injuries.[1] If 80% of this burden falls on contractors (based on their percentage of work performed), approximately 2.4 million contractor fatalities occur annually—an unacceptable number by any standard. *Safety and Health* (2023) reported that these figures demonstrate a more than 5% increase since 2015.[2]

Takala et al. (2024) further analyzed the 2019 ILO data to generate "global-, regional- and country-level estimates on the work-related burden of diseases and accidents."[3] Some of their findings are highlighted:

- The global work-related death burden increased by 4% between 2015 and 2019, rising from 2.78 million to 2.9 million deaths.
- DALY (disability adjusted life years) attributable to work were 180 million globally, an increase of 47% from 2015 to 2019.
- The economic loss from the work-related diseases was 5.8% of the global GDP.

The ILO also identified vast disparities in HSE performance between developed and developing nations:

- Fatality rates in the developed world were about half the rates of Central and Eastern Europe, China, and India.
- Latin America and the Caribbean experience even higher fatality rates.
- The Middle East and much of Asia (excluding China and India) saw fatality rates nearly four times those of industrialized nations.

DOI: 10.1201/9781003728467-1

- Some hazardous jobs were 10 to 100 times riskier in developing nations than in developed nations.
- Construction jobs in developing nations were estimated to be 10 times more dangerous than equivalent jobs in developed nations.[4]

The ILO (2023) also identified the top three contributors to workplace disease-related fatalities:

- Circulatory diseases
- Cancers
- Respiratory diseases

These contributors account for over 75% of work-related disease fatalities. Agriculture alone accounts for ~30% of workplace fatalities. Agriculture, forestry, fishing, and manufacturing collectively account for 200,000 fatalities annually, 63% of total workplace fatalities (Figure 1.1).

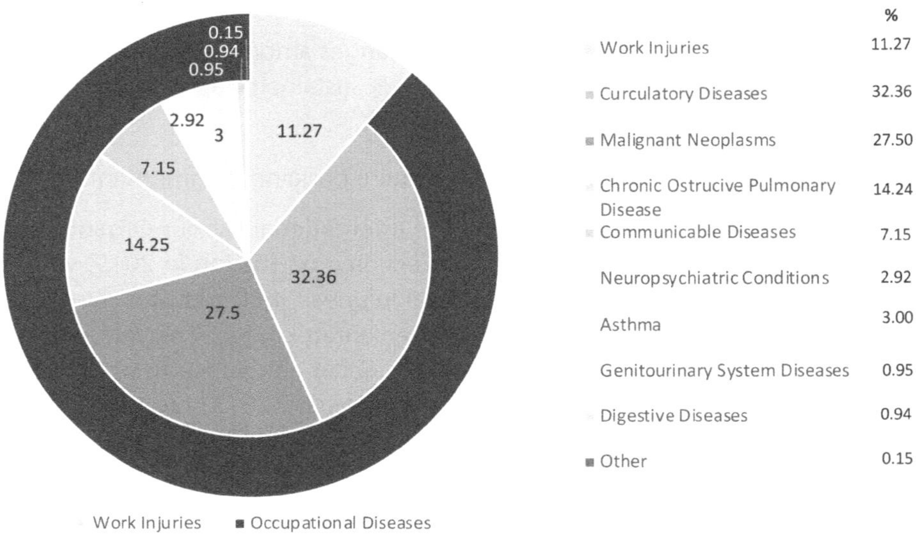

Figure 1.1: Leading occupational risk factors and total number of attributable fatalities.[5] Source: RoSPA (2020).[6]

RoSPA (2020) estimated that about 2.78 million workplace fatalities occur annually due to accidents and diseases globally, as outlined in Table 1.1.

Table 1.1: A Categorization of Total Workplace-Related Fatalities by Percent and Aggregate Number in Asia, Europe, Africa, and the Rest of the World in 2020

Region	Percent of Workplace Fatalities	Number of Fatalities (Million)
Asia	66%	1.8348
Europe	12%	0.3336
Africa	12%	0.3336
Rest of the World	10%	0.2780
Total Workplace-Related Fatalities		2.78 million

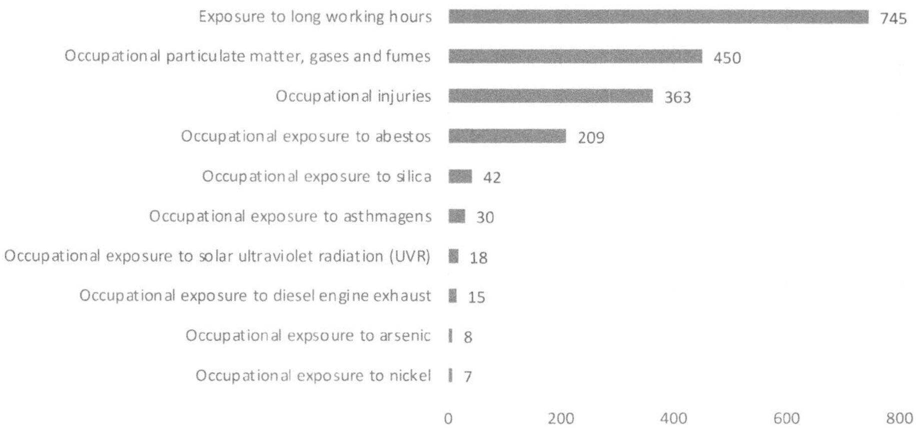

Figure 1.2: Leading global occupational risk factors and their estimated total number of attributable fatalities (thousands) in 2016.[7]

The ILO (2023) highlighted exposures to hazards and related fatalities, as shown in Figure 1.2.

Beyond fatalities, the burden placed on our healthcare systems to treat workplace injuries and illnesses is astronomical. RoSPA (2020) cited the EU-OSHA's estimate that "work-related ill health and injury cost the EU around 3.3% of its GDP a year, or 476 billion euros." The estimated direct and indirect costs of workplace accidents are provided in Table 1.2.

Table 1.2: The Estimated Annual Direct and Indirect Costs of Workplace Accidents in Australia, the United States, and Globally in 2020

Region	Percent of GDP	Annual Cost
Australia	5.9%	AU $57.5 billion
United States	1.8%	US $250 billion
Globally	1.8–6.0% of country GDP with an average of 4%	$2.8 trillion

Clearly, workplace injuries and illnesses create not only human tragedy but also an overwhelming financial and economic burden.

Can we do better? *Absolutely.*

We believe that greater awareness of these devastating numbers can help drive a collective effort to protect our workforce and save lives. To start, we must recognize the challenges and identify practical solutions. The subsequent step is to use the tools and processes available to prevent workplace injuries, illnesses, and fatalities.

To be clear, employers do not *want* workers to be harmed in the workplace. Injuries, illnesses, and deaths are costly—not only to an employer's wallet but also to their reputation. Keeping workers healthy and safe is in every employer's best interest. Today, with heavier emphasis on environmental, social, and governance (ESG) performance, organizations are under increased scrutiny and pressure to ensure a safe work environment.

Failure to meet these expectations can result in:

- Severe regulatory penalties, including the loss of operating privileges.
- Stakeholder backlash, including removal from preferred client and/or supplier lists.
- Shareholder disapproval and penalties, especially from highly socially conscious Gen Y and Gen Z investors.
- High workforce turnover—nobody wants to work in an unsafe or uncaring work environment.
- Difficulty attracting and retaining the best and brightest talent. Disciplinary actions and terminations at all levels—from frontline supervisors to executives; no organizational rank is completely safe from consequence.

Despite these imposed pressures and penalties, global HSE performance remains dismal. It's deeply troubling to us that many leaders appear to accept the current incidence of work-related injuries and illnesses, instead of aiming to pioneer solutions.

In 2020–2021, the world united against the COVID-19 virus as it threatened to wreak havoc globally. *So why can't we unite to tackle workplace fatalities and injuries*—also a global and tragically impactful issue?

For those who lose loved ones to a work-related fatality, the common experiences are heartbreaking:

- Immense grief and sorrow.
- Loss of income from a trusted breadwinner.
- Anger toward the employer—*They could have done more!*
- A sense of being disregarded and forgotten as business operations continue.
- Regret—*Could I, or would I, have done something differently had I known the risks?*
- Dramatic and immediate changes in the family dynamic (e.g., lifestyle, behavior, relationships).
- Changes in social circles, loss of friendships, and social support systems.
- Development of chronic medical conditions, including post-traumatic stress disorder (PTSD), major depressive disorder (MDD), and prolonged grief disorder (PGD).[8]

The vast majority of us who have worked in the industrial sectors know of a coworker who died in the workplace. How many of us have truly recognized or learned from their families' pain and suffering?

But it's not all doom and gloom. We *have* made progress, and the current work environment is better than it used to be—in many cases. Moving forward, the future can be brighter. Together, we can improve contractor HSE performance and significantly reduce workplace fatalities and injuries. We believe the tools and processes offered in this handbook are practical and designed for real-world application. They are customizable to any organization and for any industry.

When organizations and contractors collaborate for safety—sharing both responsibility and accountability, transformational safety culture becomes possible.

Let us commit to continued learning, cohesive action, and ultimately protecting one another.

Notes

1. International Labour Organization. 2023. *A Call for Safer and Healthier Working Environments: Global Estimates of Work-Related Accidents and Diseases 2019*. Geneva, Switzerland: International Labour Organization.
2. International Labour Organization. 2023. *A Call for Safer and Healthier Working Environments*.
3. Takala, J., Hämäläinen, P., Sauni, R., Nygård, C. H., Gagliardi, D., & Neupane, S. 2024. Global-, Regional- and Country-Level Estimates of the Work-Related Burden of Diseases and Accidents in 2019. *Scandinavian Journal of Work Environment & Health* 50(2), 73–82. https://doi.org/10.5271/sjweh.4132. Retrieved July 24, 2025, from the NIH National Library of Medicine.
4. International Labour Organization. 2023. *A Call for Safer and Healthier Working Environments*.
5. International Labour Organization. 2023. *A Call for Safer and Healthier Working Environments*.
6. RoSPA (2020). Industrial Safety. *Occupational Safety & Health Journal*, 50(1); Special section p. 07–08. Retrieved June 5, 2021, from EBSCOHost database.
7. International Labour Organization. 2023. *A Call for Safer and Healthier Working Environments*.
8. Matthews, L. R., Quinlan, M. G., & Bohle, P. 2019. Posttraumatic Stress Disorder, Depression, and Prolonged Grief Disorder in Families Bereaved by a Traumatic Workplace Death: The Need for Satisfactory Information and Support. *Frontiers in Psychiatry* 10:609. Accessed September 7, 2025. https://doi.org/10.3389/fpsyt.2019.00609

2 A LEADERSHIP MINDSET CHANGE

Leadership Mindsets and Behaviors Towards CSM

Dr. Chit Lutchman: As a frequent presenter at international HSE conferences and workshops, I have spent more than a decade presenting on contractor safety management (CSM) globally. In preparing for such events, presenters typically define the target audience and participants. I seek to include general managers, vice presidents, operations personnel, maintenance and procurement leaders, frontline supervisors, and contractors in CSM presentations and workshop training. However, actual attendance is often skewed toward those in lower-level roles whose CSM responsibilities are tangential to their core organizational duties. Contractors are rarely present at these conferences and workshops. Many participants simply attend to better understand the contractor management process in general.

Common themes and observations from interactions with these participants include several recurring leadership mindsets and behaviors that may actually hinder CSM improvements. These include:

- My organization's CSM is effective and requires no additional attention or focus.
- CSM is an economic and financial decision with lip-service consideration for HSE focus and attention.
- Senior leadership isn't directly responsible for CSM; it's the frontline supervisors and operational leaders who are both responsible and accountable.
- Any incident occurring during the contractor's work is the contractor's responsibility.
- Contractor prequalification is sufficient to measure a contractor's ability to perform work safely, reliably, and efficiently.
- CSM occurs at the frontline, and events are the outcomes of unforeseen hazards, risks, and unpredictable circumstances.

Let's explore these leadership mindsets in more detail to better understand their implications.

My Organization's CSM Is Effective and Requires No Additional Attention or Focus

This perception often arises after periods without serious incidents. Often, leaders become complacent, failing to track relevant leading indicators that could be precursors to serious incidents and fatalities. These perceptions may include:

- We have established an effective contractor and supplier management protocol that adequately defines our expectations and requirements.
- We effectively communicate our expectations and requirements to our contractors.
- Contractor work practices and processes verifications are completed consistently and effectively. Our documented cross-checks during pre-mobilization activities ensure contractors can perform the work safely, reliably, and efficiently.

DOI: 10.1201/9781003728467-2

- We effectively verify contractor workforce training and competency records before work is performed.
- Our contractor work practices and procedures evaluations ensure the contractor meets the requirements and expectations detailed in the scope of work (SOW).
- Near-miss events within the contractor workforce are not our immediate concern. Events at another client's worksite do not concern us since our CSM is superior and effective.
- Our performance management indicators are effective and ensure we are proactive in mitigating events.

> **NOTE:** Overreliance on incident rates, recordable injury frequencies (RIFs) and other lagging indicators may allow minor operational deviations to precipitate major accidents.

To contextualize the potential effect of overconfidence, let's highlight a historical event.

The National Aeronautics and Space Administration (NASA) is a US federal agency regarded for its deeply rooted safety culture. It boasts a 98.5% space shuttle mission success rate.[1]

On January 28, 1986, NASA's spacecraft STS-51L, Challenger, exploded 73 seconds into its flight, killing its seven-astronaut crew.[2] The internationally televised incident was observed by millions of people, illustrating the potentially disastrous consequences of organizational complacency regarding HSE and CSM.

Decision-makers at NASA and its primary contractor, Morton Thiokol, Inc., failed to properly address engineers' concerns about the solid rocket booster (SRB) rubber O-rings—intended to prevent leaks around the SRBs.[3] Testing from 1977 showed flaws in the O-ring design, and NASA engineers opined that the O-rings should be equipped with shims to reduce the risk of combustion gas erosion. Post-flight evaluation of the O-rings in 1981, 1984, and 1985 all showed evidence of erosion, but not enough to compromise the flight.[4] Near miss events, indeed.

However, NASA—steered by its innate ambition—targeted an increase in flights from 15 per year in 1986 to 24 per year by 1990 with relatively consistent funding (Waldrop, 1986).[5] *Cheaper, better, and faster* was the goal.

The morning of the Challenger's launch was particularly cold. The ambient temperature reduced the compliance of already compromised O-rings, contributing to their failure and the horrific loss of seven astronauts.[6] According to the 1986 US Congressional Budget Office, the accident's immediate operational costs to US taxpayers were likely to exceed several billion dollars.[7] The price paid by the astronauts and their loved ones is immeasurable.

Despite clear evidence indicating that the O-rings could fail, organizational success overshadowed the need for ongoing rigorous safety assessments and responses to a seemingly minor issue. Complacency fostered insufficient risk mitigation and permitted poor contractor accountability, ultimately resulting in catastrophe.

This was NASA's first failed shuttle launch. This disaster stresses the necessity of a vigilant safety culture—no matter how flawless the current system appears—and for all contractor management practices to prevent organizational blind spots from producing fatal consequences.

CSM Is an Economic and Financial Decision

For many organizations—particularly in regions such as the Middle East and Asia—cost (rather than HSE performance) is the primary driver of the contractor selection process. Organizations tend to hire underqualified low-cost bidders, often at the expense of safety performance. Even weak HSE performance records may be overlooked if the price is right.

Equally problematic is contractor tendency to undervalue the benefits of an effective safety management system, given the financial outlays required to establish an HSE management system. The summarized dialogue below, during a CSM workshop, highlights economic and financial variables as principal determinants in CSM.

Dr. Chit Lutchman: I recall a telling conversation with the owner of a contracting company at a CSM workshop in a major industrial region:

Contracting Company Owner: Why should I invest $1 million in a safety management system? If a contract worker dies while working on my project, I can go to their home country, give their family $50,000, and they're satisfied.

Aghast and at a loss for words, I recognized the participant was addressing CSM as a purely financial decision. I had to be tactful in my response.

Dr. Chit Lutchman: How do you want to be remembered by this family? Do you want this person's family to revere you as a savior, a provider of opportunity? Or would you prefer to be remembered as the man who robbed them of their loved one?

The brevity of the owner's reflection was telling. It illustrated the deeply embedded reality of the situation: *cheaper, better, and faster* operations are prioritized by both owner and contractor organizations, often undermining the value of a worker's life—especially in high-risk work environments.

Senior Leadership Is Not Responsible for CSM

Classically, most leaders point a finger at frontline supervision when asked about CSM. While senior leaders may acknowledge their ultimate accountability for contractor safety, they rarely take ownership of the responsibility.

This attitude, however, has begun to shift—particularly in the energy industry.

Dr. Chit Lutchman: Over my 30-year career, I've witnessed a growing recognition and acceptance of shared responsibility for contractor safety among leaders. Gradually, we've shifted from ***us and them*** *(owners and* *contractors)* to ***us*** *(a joint owner and contractor ownership),* which indicates meaningful progress in a quest for safe, reliable, and efficient contractor performance.

A compelling demonstration is the resignation of Mr. Mark Little, former CEO of Suncor Energy (Canada's largest oil and gas company), in 2022 after several contractor fatalities over a seven-year period. While the organization implemented many programs and processes to improve contractor safety performance, success was elusive. So, Mr. Little decided to walk away and allow another leader the opportunity to generate necessary change.

Elliot Investment Management, an American management firm and activist fund, raised concerns about Suncor's board of directors and management regarding these fatalities and other operational challenges. Seskus (2022) pointed to Eight Capital analyst Phil Skolnick, who advised that "Suncor requires 'real changes,' not just a new CEO" in response to 13 workplace fatalities since 2014.[8] The frequency of fatalities was described as "something that we have never seen in the 25 years of covering the sector." According to Skolnick, "The issue isn't a one-person situation." He wrote further, "this is about the corporate culture where accidental deaths have plagued the company even prior to Mr. Little's tenure as CEO (which) began in 2019."[9]

In 2022, Suncor's interim CEO acknowledged the company's need for decisive action, stating that Suncor "must start putting solutions in place to address workplace safety problems." Following an independent safety assessment, he emphasized that the organization was "clear on what we need to do to improve our safety performance," urging the execution of improvement measures (Wilson, 2024).[10] Suncor has since reported that 2023 marked its strongest safety performance in nearly a decade, with no life-altering or life-threatening injuries for the first time since 2015 and nearly a 50% reduction in lost-time incidents (LTIs).[11]

This fragile reprieve may be the fine line between luck and genuine improvement. Only time will tell if there have been sufficient structural and management system changes—and senior leadership ownership and accountability—to sustain improvements in contractor safety performance.

Contractor Safety Is the Contractor's Responsibility

In many organizations, leaders treat contractor safety events as a contractor failure. Events may be investigated by the organization and/or contractor, and root and contributing causes may be identified to define and implement corrective actions. In many instances, however, owners fail to look inward to better understand how internal practices may have contributed to the event.

Today, owners continue to select contractors with a mindset of getting contracted work done *cheaper, better, and faster.* Again, theoretically compatible goals that often antagonize each other in the real world. When incidents do occur, contractors are unceremoniously dismissed or instructed to fix internal problems at the contractor's expense with little regard for contractor earning margins.

Owners may overlook their own internal unsafe practices including:

- Unrealistic schedules

- Cost prioritization, which reduces or eliminates safety and training budgets
- Inadequate oversight or support

A relatively recent historical example of the potential disastrous outcome of over-delegating CSM to contractors is the catastrophic BP Texas City refinery explosion of 2005. The incident claimed the lives of 15 workers (all contractors) and injured 180. Most of the fatalities of this incident were of workers stationed in temporary contractor trailers sited far too close to hazardous operational equipment.[12]

Subsequent investigations identified weak process-safety leadership—with 439 OSHA citations—and overreliance on lagging indicators (such as injury frequencies and rates) as contributors to the event in the setting of insufficient contractor oversight.[13]

Contractors were often unaware of critical hazards and were allowed to operate from trailers located within blast zones without host-controlled protections.

The US Chemical Safety Board (CSB) opined that had BP maintained appropriate oversight and facilitated the necessary safety training, they may have identified hazardous operational conditions and may have enforced safe trailer distances, which may have saved the 15 lives lost.[14]

BP failed to ensure its contractors adhered to critical safety processes.

OSHA and BP settled penalties of over $13 million USD and mandated corrective actions, underlining that contracting work does not contract all safety responsibility.[15]

The Texas City tragedy shows us that the absence of steadfast owner oversight, integrated and collaborative safety systems, and enforced safety controls puts contractors directly at risk—whether the risk is already identified or not.

Organizational application of a more collaborative approach, where owners and contractors jointly identify and resolve systemic issues, can yield mutually beneficial and sustainable performance improvements.

Prequalification Is a Sufficient Measure of Contractor Capability

Fundamentally, many owners misunderstand the purpose and limits of contractor prequalification—particularly when performed by a third-party prequalification service provider. Owner representatives tend to accept green-flagged (*pre*qualified) contractors as *qualified* to perform the expected work safely, reliably, and efficiently.

A *prequalified* designation can be misleading, as contractors may ensure they are prequalified within a third-party prequalification organization database with the owner having no knowledge of the database contents beyond its insurance certification. The contractor's principal goal is to stay green-flagged—prequalified in a database—maximizing the chances of being selected for a project. In many cases, contractors are prequalified without the owner or prequalification service provider ever verifying:

- Documented safe work practices and procedures—content, quality, and usability

- Safety practices
- Job-specific competency
- Individual worker competency and team collective competencies
- Job and site-specific performance capability—tools, equipment, and machinery availability (not to mention their maintenance)

Ultimately, contractors want to appear ready for business. As such, owner representatives cannot abandon their vetting responsibilities and obligations by blindly relying on third-party systems.

Dr. Chit Lutchman: During my earlier career, as an employee tasked with selecting a prequalification service provider for my employer, I was able to receive a green flag in the prequalification service provider database by uploading blank documents. When confronted by the service provider about not being selected—and having explained my actions and observations—I was advised by the service provider that verification was the owner's responsibility. I will concede that improvements have been made over time, with service providers checking to see if documented uploads have content. Unfortunately, challenges continue to exist. In many instances, such verifications are still performed as a rubber-stamping activity by administrative non-subject matter expert personnel.

Today—more than 15 years later—as I work with clients across North America developing CSM standards, tools, and processes, many owners continue to use prequalified contractors with complete reliance on a prequalification service provider to perform all verifications in the qualification processes. In many cases, the owner remains oblivious contractor deficiencies for the SOW to be performed. As such, contractor injuries, illnesses, and fatalities continue to be prominent events in our workplaces.

Owners must further qualify contractors with more robust checks and verifications to confirm a contractor is truly qualified and competent to perform the desired work safely, reliably, and efficiently. Failure to do so jeopardizes both contractor safety and organizational integrity. We continue to run the risk of repeating major events while already possessing the tools to avoid them entirely.

CSM Is a Frontline Activity and Events Are Unpredictable

Most leaders will argue that much of CSM occurs at the frontline where workers are faced with tangible hazards and risks. Such leaders also believe that hazard and risk mitigation is the responsibility of the frontline workforce. However, we believe CSM is an integrated responsibility that requires baseline operational risk profiling and continuous field data collection and analysis to predict and mitigate safety events.

Effective CSM is a joint leadership and frontline activity as follows:

- The impact of leadership engagement and visibility in the execution of work is among the most powerful determinants of contractor safety performance. Leaders can:
    - Clearly define safety expectations, integrating them into qualification criteria.

    - Deter unsafe behaviors with effective leadership visibility, supervision, oversight, audits, and utilization of performance metrics.
    - Provide the necessary tools to support safer work.
    - Encourage the reporting of near misses and incidents through non-punitive corrective actions.
- Effective leadership sets the expectations and establishes the requirements for safe, reliable, and efficient (excellence) business performance from all workers, thereby establishing the precursors of the organizational safety culture.
- Effective hazard identification, as well as risk assessment and mitigation, are the outcome of an integrated, competent, and experienced stakeholder group (leaders and frontline).
- Many studies estimate that 90–95% of workplace events are predictable and are therefore avoidable.

To further illustrate this, let's look at a historically recurring project.

Every four years, a major city braces for the arrival of the world's premier athletes—and humanity's attention. Hosting the Olympic Games usually entails the frantic construction of an Olympic Village and a world-class stadium fit for international spectacle.

In 2012, London was set to host the Olympics. The plan was to repurpose a 200-hectare brownfield site in Stratford, East London, into the Olympic Village and Stratford Stadium—a massive structure capable of seating 80,000 people. Not only was the worksite an industrial dump, wrought with hidden hazards, but the project was highly complex with a strict timeline requiring approximately 65,000 workers (Waterman et al., 2011).[16]

The United Kingdom's Olympic Delivery Authority (ODA) felt that the notion of *expected* worker fatalities during Olympic Village construction was unacceptable and sought to set a new safety standard. According to the Institution of Civil Engineers Report (2011), "The ODA accepted fully that it had a moral obligation to minimise harm to its workforce," including its employees and contractors.[17]

The ODA meticulously planned the project, intertwining health and safety into every component of the blueprint. They identified key stakeholders early on who could provide leadership in mitigating hazards before they caused harm to the workforce. They identified opportunities for optimizing the health of their workforce, partnering with occupational health teams—including physicians, nurses, physiotherapists, and industrial hygienists—to provide medical evaluations for safety critical workers, health surveillance, drop-in treatments, and campaigns on wellbeing. Notably, after identifying internally that many of their workers were obese and maintained poor diets, they even offered a low-cost, healthy porridge breakfast on the worksites to circumvent increased midday accidents because of skipped breakfasts after heavy, fatty dinners the night before (Waterman, 2013).[18]

From top-level executives down, the ODA created unanimous buy-in to a rigorous and collaborative safety culture, which not only facilitated on-time and under-budget completion of the project but set a new benchmark for operational excellence in Olympic venue construction. After more than 60 million hours of work, "the onsite accident frequency rate was 0.17 per 100,000 hours," which made this construction site safer than the UK's all-industry average.[19] Not to mention, for the first time in Olympic Village construction history, no workers died.

Leadership is a critical determinant of contractor safety performance. Organizations successful in CSM demonstrate a structured approach to shaping safe, reliable, and efficient work environments—where leaders at executive, mid-level, and frontline positions accept responsibility for CSM. Still, far too many leaders continue to view CSM as a box to check and a costly endeavor rather than an integral component of organizational function.

We must pursue fundamental shifts: from compliance to commitment, and from delegation to connection.

Notes

1. National Aeronautics and Space Administration. 2025. Space Shuttle. Accessed August 28, 2025. https://www.nasa.gov/space-shuttle/
2. Encyclopaedia Britannica. 2025. Challenger Disaster. Accessed August 28, 2025. https://www.britannica.com/event/Challenger-disaster
3. Lach, C. L. 1989. *O-Ring Sealing Verification for the Space Shuttle Redesign Solid Rocket Motor* (NASA-TM-101591). Hampton, VA: National Aeronautics and Space Administration, Langley Research Center. Accessed August 28, 2025. https://ntrs.nasa.gov/api/citations/19890016897/downloads/19890016897.pdf
4. Mallak, C. T. 2018. Death on Orbit: Extreme Environmental Conditions and the Deaths of American Astronauts. *Academic Forensic Pathology* 8(1):112–118. Accessed August 28, 2025. https://doi.org/10.23907/2018.008
5. Waldrop, M. M. 1986. After Challenger: Painful Choices. *Science* 232(4756): 1335–1337. Accessed August 28, 2025. https://www.science.org/doi/10.1126/science.232.4756.1335
6. Mallak, Death on Orbit, 2018.
7. Congressional Budget Office. 1986. Budget Effects of the Challenger Accident (99th Cong., 2d Sess., S. Doc. No. 8, Part A). Washington, DC: U.S. Government Printing Office. Accessed August 28, 2025. https://www.cbo.gov/sites/default/files/99th-congress-1985-1986/reports/doc08a_0.pdf
8. Seskus, T. 2022. Suncor Boss Mark Little Stepping Down 'Effective Immediately'. Accessed March 3, 2025. https://www.cbc.ca/news/canada/calgary/mark-little-leaves-suncor-1.6515578
9. Seskus, Suncor Boss Mark Little Stepping Down, 2022.
10. Wilson, J. 2024. Suncor Claims 2023 is its Safest Year, Canadian Occupational Safety. Accessed August 18, 2024. https://www.thesafetymag.com/ca/topics/safety-and-ppe/suncor-claims-2023-is-its-safest-year/478534
11. Wilson, Suncor Claims 2023 is its Safest Year, 2024.
12. U.S. Chemical Safety and Hazard Investigation Board. 2007. Refinery Explosion and Fire at BP Texas City Refinery – Final Investigation Report (Report No. 2005-04-I-TX). Washington, DC: U.S. Chemical Safety and Hazard Investigation

Board. Accessed September 15, 2025. https://www.csb.gov/assets/1/20/csbfinalreportbp.pdf

13. U.S. Department of Labor. 2012. BP Agrees to Pay more than $13 Million and Abate Violations in Settlement Agreement with US Department of Labor. Release 12-1383-NAT. Washington, DC: U.S. Department of Labor, OSHA. Accessed September 15, 2025. https://www.dol.gov/newsroom/releases/osha/osha20120712
14. U.S. Chemical Safety and Hazard Investigation Board. 2007. Refinery Explosion and Fire at BP Texas City Refinery.
15. U.S. Department of Labor. 2012. BP Agrees to Pay more than $13 Million and Abate Violations.
16. Waterman, L. et al. 2011. Delivering London 2012: Health and Safety. *Proceedings of the Institution of Civil Engineers — Civil Engineering* 164(CE5):46–54. Accessed [insert access date]. https://doi.org/10.1680/cien.2011.164.5.46
17. Waterman et al., Delivering London 2012: Health and Safety, 2011.
18. Olympic Delivery Authority. 2013. Olympic Delivery Authority: Annual Report and Accounts 2012–13. London: Olympic Delivery Authority. Accessed September 9, 2025. https://assets.publishing.service.gov.uk/media/5a7b8b0ee5274a7318f570/ODA_Annual_Report___Accounts_2012-2013.pdf
19. Olympic Delivery Authority, Annual Report and Accounts 2012–13, 2013.

3 CONTRACTOR AND SUPPLIER SAFETY MANAGEMENT PROCESS

CSM Overview Process

In this chapter, we'll introduce a globally accepted, multi-industry Contractor and Supplier Safety Management Process. In its simplistic form, the Contractor and Supplier Safety Management Process comprises the stages shown in Figure 3.1.

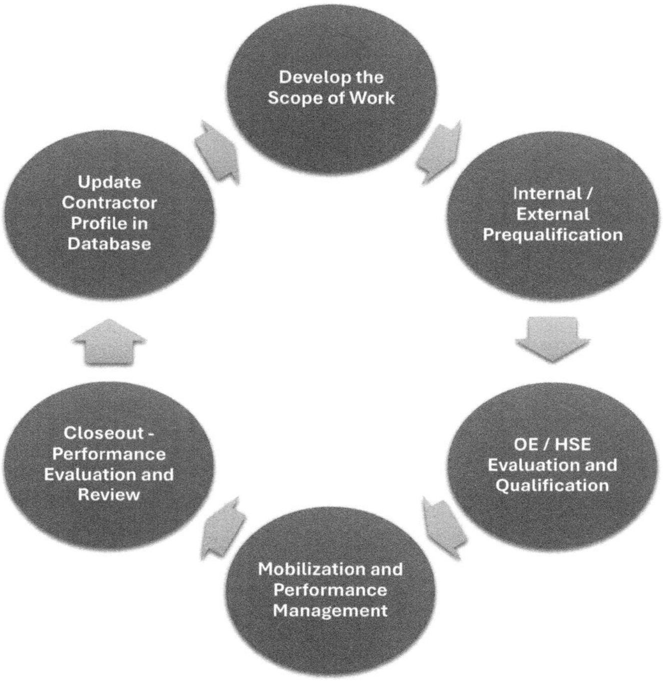

Figure 3.1: Stages in the Contractor Safety Management Process (CSM Cycle).
Source: OEMS Erudite Corp. (2025).

Detailed CSM Process

Figure 3.2 provides detailed stages in an effective Contractor and Supplier Safety Management Process. Most organizations implement incomplete variants of the detailed process, often overlooking or underutilizing high-value opportunities for improving contractor HSE performance. Common process flaws include:

- Many owners do not require contractors to develop an HSE Plan for assessment in the scope of work (SOW) evaluation process.
- Many owners forego an owner-led contractor HSE Alignment Session to ensure the contracting workforce is aware of the requirements and expectations during work execution.
    - The HSE Alignment Session is a more detailed process than an orientation that highlights known hazards and risks associated with the work.

DOI: 10.1201/9781003728467-3

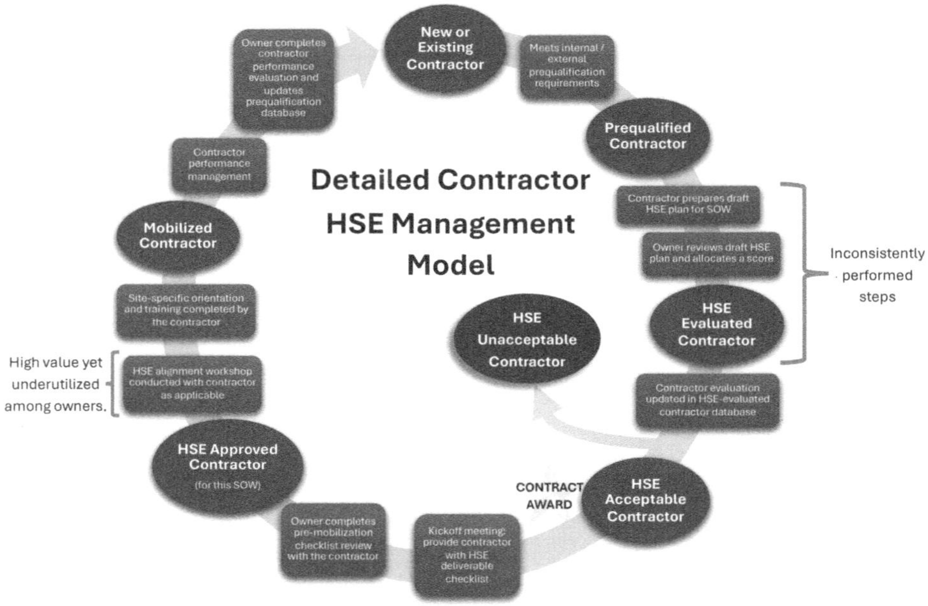

Figure 3.2: Detailed steps in the contractor safety management process.
Source: OEMS Erudite Corp. (2025).

Contractors and Suppliers—Differentiated

Why Contractors and Suppliers Are Handled Differently

To clearly understand the CSM process, we must distinguish contractors and suppliers, as organizations typically manage them differently.

Contractors	Suppliers
• Contractors are generally classified as organizations or individuals providing services to the client or owner. • Contractors are generally required to maintain their own documented practices and procedures to execute work, meeting or exceeding owner requirements. • Contractors may provide a transient workforce to the owner.	• Suppliers tend to provide materials and goods to the client or owner and generally do not perform hands-on work at the owner's sites. • Suppliers may be required to install equipment and machinery at the owner's sites and often work under their own HSE management systems and documented practices and processes. • Owners may provide oversight to the supplier during the delivery and installation of equipment and machinery.

• Depending on the SOW hazards and risks, owner supervision and control may vary as follows:
  ○ High risk—full owner supervision and control.
  ○ Moderate risk—intermittent owner supervision with full contractor supervision and control.
  ○ Low risk—no owner supervision with full contractor supervision and control.
• Contractors may be fully supervised by owners and may be required to retain onsite HSE resources to monitor HSE performance during work execution.
• Contractors are often hired by owners to perform *3-D* work as follows:
  ○ **D**irty
  ○ **D**ifficult
  ○ **D**angerous
• Depending on the SOW, the contractor's workforce may vary, up to several thousand people and for short to extended periods (where the contract worker functions as an employee).
	• Supplier workforces may be longer-term than the contractor's workforce.
	• Suppliers are generally experts in the execution of their work and are well-known for the supplies they provide.
	• The supplier's primary focus is on quality verification of delivered equipment, machinery, materials and supplies.
	• Work is completed in short periods by a small number of very experienced and competent workers.

Are Contractors and Suppliers Subject to the Same Process?

Most contractors and suppliers must satisfy the same HSE requirements and expectations defined by the owner organization. For suppliers, however, the criteria may be less stringent since they may not perform the same level of hands-on work at the owner's sites as do contractors. Nevertheless, both contractors and suppliers often undergo the same qualification process.

Simplified Contractor and Supplier Safety Management Process

Overview

The simplified Contractor and Supplier Safety Management Process provides a summary of the major activities expected when managing contractors and suppliers.

- Required activities are generally completed within their respective stage of the process.

- All stages are addressed in an effective Contractor and Supplier Safety Management Process.

Development of the Scope of Work (SOW)

SOW development is the first opportunity to ensure excellence in work execution. When developing the SOW, it is important that the right owner stakeholders are available to participate.

- The SOW is a documented list of contractor and/or supplier deliverables.
- In SOW development, HSE and risk management stakeholders play a critical role in identifying and documenting hazards, risks, and statutory compliance requirements (including contractor wellbeing) as well as defining mitigation actions required by the contractor or supplier.
- Clear documentation of known hazards, risks, and statutory compliance requirements must be communicated in the SOW documentation. Contractors and suppliers address the hazards and risks associated with the work in their documented bid response.
    - A contractor HSE management plan and bridging documentation are required by some owners.
- The SOW definition may include the following deliverables components and expectations:
    - Engineering and design
    - Quality—reliability and integrity
    - HSE
    - Other—per relevant stakeholder groups
    - Commercial—Cost of service. To be submitted separately from other deliverables to the procurement team.

> **NOTE:** Commercial expectations are generally excluded from the SOW definition and are communicated to the contractor or supplier to send to the procurement/supply chain department under separate cover.

The HSE SOW and expectations from the contractor are best demonstrated in Figure 3.3.

SOW deliverables components vary based on the project's magnitude. For most major projects, contractors are likely required to deliver the following:

- Work aligned with the owner's CSM expectations and requirements. See Appendix 1—Contractor Safety Management Standard/Protocol for sample expectations and requirements.
- An Environmental Management Plan, which may include the following components:
    - Water management plan
    - Waste management plan
    - Spills prevention and response plan
    - Regulatory compliance deliverables
    - Other environmental plans/procedures

- A Health, Safety, and Security Management Plan, which should include the following components:
    - Safety plan and procedures
    - Occupational health plan and procedures
    - Emergency response plan
    - Security plan

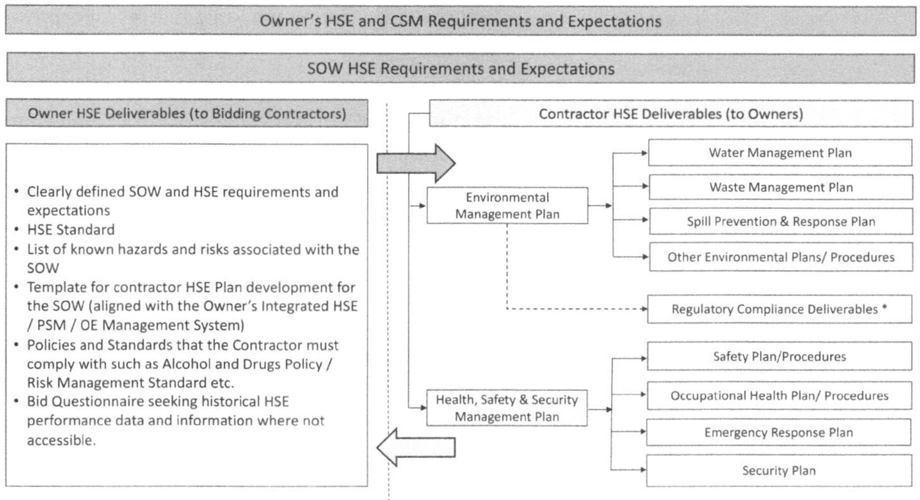

Figure 3.3: Typical HSSE/HSE/EH&S scope of work components for major projects.
Source: OEMS Erudite Corp. (2025).
* Regulatory compliance deliverables may vary based on the owner's desire to meet or exceed legal requirements and expectations.

Internal/External Prequalification

Prequalification is typically the first step for contractors and suppliers to demonstrate their capability to meet an owner's requirements and expectations for excellence in the delivery of work. Contractors and suppliers usually present the owner with verifiable evidence of safe and reliable work performed for other clients.

Demonstrated verifiable information required by the owner generally includes:

- Historical HSE performance records (e.g., incident rates, fatalities, major accidents and injuries, spills and environmental incidents)
- Insurance documentation confirming insurability and compliance with the owner's insurance coverage requirements
- Documented standards, procedures, tools, and templates
- References and/or testimonials from former clients
- Training and competency assurance records of the participating workforce
- Liability insurance
- Other SOW-specific components as required by the owner organization

An owner representative reviews the prequalification documents against predefined minimum expectations to determine a contractor or supplier prequalification status. Owners generally limit the number of reviewable bids on an SOW to value-maximize evaluations.

Prequalification may be conducted by the owner internally using in-house resources or externally using a contracted service provider. The attributes of internal and external prequalification are summarized below.

Internal Prequalification	External Prequalification
• Performed by the owner organization. • Time-consuming, limiting the number of contractors and suppliers within the prequalification database. • May be difficult to standardize against industry peers, but generally more robust and effective than external prequalification by utilizing owner-customized criteria. • Trusting contractor and supplier relationships with the owner can be established and maintained. • Estimated costs of performing the work safely, reliably, and efficiently are likely more accurate. • Sometimes performed after suppliers refuse to be prequalified externally. • High vested interest in getting prequalification done right.	• Performed by an external service provider organization. • Efficient, given the use of industry standardized criteria and more automated processes. Pricing may be more competitive to owners. • Contractors and suppliers are charged by the prequalification service provider to become prequalified in its database. These fees may be imparted on the owner within the bid. • Less aligned with the specific risks and hazards associated with the SOW. • Contractors and suppliers can outsource the supply of required documentation to the external prequalification agency, possibly having never seen the documents themselves—a checkbox exercise to become prequalified. • Low vested interest, as external prequalification platforms are financially incentivized to expand the contractor and supplier numbers in their database. Essentially, lower prequalification standards are more profitable.

HSE Evaluation and Qualification

HSE evaluation and qualification is an integral step in selecting the right contractor or supplier. Many organizations employ a weighted scoring process, which generally includes the following:

Evaluation Criteria	Weighted Score %
Engineering and Design	25
Quality—Reliability and Integrity	25
HSE/EHS	30
Commercial (Pricing)	10
Other Relevant Stakeholder Groups Criteria	10

NOTE: Allocated weighted scores may vary within the organization and across organizations.

- HSE evaluations may facilitate veto actions where a contractor is excluded from the selection process if they fail to meet a satisfactory HSE score, regardless of their scores in other disciplines.
- Document (i.e., standards, procedures, HSE manuals, etc.) reviews and assessments are used to determine contractor suitability based on the owner's SOW requirements per Appendix 2—Contractor HSE Content to be Reviewed.
- Objective scoring of a contractor's documented processes and ability to meet the owner's SOW deliverables are per Appendix 3—Sample HSE Evaluation and Scoring Process.

It is important to note the depth and breadth of evaluation and qualification is determined by the following:

- Magnitude of the SOW—the larger the SOW and associated activities and cost, the broader and deeper the evaluation and qualification process.
- Hazards and risks associated with the SOW—the more hazards and risks, the broader and deeper the evaluation and qualification process.

Owners often erroneously accept prequalified contractors as qualified without performing due diligence qualification reviews, occasionally with disastrous consequences.

Mobilization and Performance Management

Contractor mobilization and performance management is perhaps the most important stage in the Contractor and Supplier Safety Management Process. Once selected, the onus is on the owner to verify contractual obligations of people, processes, systems, and equipment to perform the SOW safely, reliably, and efficiently are met.

Lumped together in this simplified process, mobilization and performance management do not appear difficult. However, the magnitude of the owner's supervisory and oversight lift depends on a multiplicity of factors that can result in major events and incidents if not adequately addressed. Contractors' workers are vulnerable for the following reasons, among others:

Mobilization	Performance Management
• A new worksite with new rules and practices. • Workers may have to forget former practices and learn something new at the new worksite for the same work activity. • Known and unknown hazards exposures.	• Conflicting owner and contractor cultures. • Limited to no owner and contractor oversight and supervision. • Unavailable and inaccessible standards, practices, and procedures.

• Many moving parts—for example, a rig mobilization may include hundreds of truckloads of heavy equipment operating in tight spaces (e.g., narrow and/or remote roads). • New working and reporting relationships. • Young and inexperienced workers—on a new worksite, almost all workers are inexperienced.	• Inaccessible software tools and processes (e.g., incident management, inspections, and shared learning) that are HSE management system aligned. • Inadequate subcontractors performance management. • Communication and language barriers.

Closeout—Performance Evaluation and Review

Contract closeout with contractor performance evaluation and review is often performed superficially, often without the right stakeholders—if performed at all. The closeout process also includes demobilization, characterized by the following challenges and concerns:

- Contractor worker preoccupation with the next job may be distracting, potentially contributing to demobilization safety events.
- Many moving parts (e.g., over 300 truckloads of heavy equipment during rig moves) require complex coordination and can lead to safety events.
- Owner and contractor performance evaluation discrepancies may sour relationships.
- Contractor performance evaluation may be initiated after demobilization by a non-involved stakeholder—a checkbox exercise—which may strain owner and contractor relationships, particularly for future work opportunities.

Some owners, however, do a great job of sitting with the contractor before demobilization to capture learning and identify improvement opportunities. Best practices can be established (what worked well) while opportunities for improvement can be identified and improved upon before repeating the SOW.

Update Contractor Profile in Database

Updating the contractor's profile in the owner's database is often an overlooked activity. It is not uncommon to see the return to service of poorly performing contractors. Reselecting poorly performing contractors occurs because of a failure to update the contractor's profile and performance evaluation records in the owner's prequalification database.

When a similar or repeat service is required, the procurement team defaults to the last evaluated provider of similar services to select as the service provider. In some instances, a multi-service agreement (MSA) may be available for the contractor where performance is not updated in the database between services. Thus, a perpetuation of poor performance may exist.

Updating the contractor's profile in the database after each SOW provides the latest performance results that may be valuable in reaffirming or changing the contractor.

Detailed Steps in Contractor and Supplier Safety Management Process

Development of the Scope of Work (SOW)

Developing the SOW is essentially the same in both the simplified and detailed processes. Occasionally, where the SOW is unclear, the owner may seek input from its applicable competent contractor group to establish the SOW requirements in a request for information (RFI).

When the SOW is defined, and a physical worksite visit helps clarify the SOW requirements to eligible contractors, a pre-bid site visit may be required to achieve the following:

- Communicate known hazards and risks—and identify those previously unknown—associated with the SOW and any other details unique to the SOW.
    - See Appendix 4—Pre-bid Meeting Form.
- Establish plot plan requirements and site ownership and controls requirements (prime contractor status) for the SOW.
    - See Appendix 5—Assigning Ownership of the Worksite—Prime Contractor Transfer.
- Ensure all contractors clearly understand the SOW and associated hazards and risks so that they may be fully addressed in the bid response.

Prequalification—New or Existing Contractors

Both new and existing contractors are required to be prequalified per the owner's prequalification process. Owners encourage new contractors to become prequalified within the owner's database to increase the owner's choice options and to improve price and quality competition among contractors.

New and existing contractors are verified and prequalified as follows:

New Contractors	Existing Contractors
• Contractors must meet the requirements and provide documented proof that the requirements are satisfied. • Submitted documentation is verified by the prequalification service provider (internal or external) as having met the owner's requirements and expectations.	• The contractor exists within the owner's prequalification database and has provided documented proof that requirements are satisfied. • Submitted documentation is verified by the prequalification service provider (internal or external) as having met the owner's requirements and expectations.

<table>
<tr>
<td></td>
<td>

- Upon satisfactory evaluation of documented requirements, the contractor is assessed as follows:
  - Green flagged—prequalified and has met the requirements.
  - Red flagged (not prequalified)—has not provided adequate evidence to meet the owner's requirements.
  - Yellow flagged—does not meet all the owner's requirements and has been requested to provide additional information.

</td>
<td>

- Required information may be expired and may require updating in the database.
- May be adjusted to red, yellow, or green flagged based on information reviews within the database.

</td>
</tr>
<tr>
<td>Meets Internal/ External Prequalification Requirements</td>
<td colspan="2">

Prior to being invited to bid on a SOW, the owner's procurement team or SOW representatives will verify that the contractor is prequalified (green flagged) within the prequalification database.

- Occasionally, a yellow flagged contractor may be invited to bid if deficiencies can be addressed prior to a contract award, or where the deficiency does not present a risk to safe, reliable, and efficient SOW delivery.
- Occasionally, where a single source contractor is required, prequalification may be waived. See Appendix 6—Sample Prequalification Waiver Form.
- Red flagged contractors are generally not invited to bid on a SOW. However, in scenarios where a red flagged contractor is the only service provider available, they may be invited to bid with owner-implemented risk mitigation actions to ensure safe, reliable, and efficient SOW delivery.

</td>
</tr>
<tr>
<td>Contractor Prepares Draft HSE Plan for Scope of Work</td>
<td colspan="2">

For large projects with complex work activities and multiple hazards and risks, the contractor will be required to provide a draft HSE Plan for the SOW. Most contractors will simply compile a list of procedures and practices into a consolidated document to fulfil the requirement.

They do so for two reasons:

- Bid success is not guaranteed, and preparing a SOW-specific HSE Plan is time consuming and costly.
- If the contractor provides a well-developed HSE Plan and is unsuccessful in their bid, the owner may provide their HSE Plan to the successful bidder to use.

</td>
</tr>
</table>

Owner Reviews Draft HSE Plan and Allocates a Score to Contractor

A well-developed HSE Plan gains favor by the HSE evaluation team, which does not want to review and evaluate non-core documents accompanying the bid. The team may be evaluating several contractor bids, and the task can be extensive. Non-specific information adversely impacts the evaluation process where competent owner HSE resources are limited. In addition, scoring consistency may be compromised if HSE evaluation personnel must work independently to manage volume.

HSE Evaluated Contractor

Per Appendix 3 (Sample HSE Evaluation and Scoring Process), each bidding contractor is evaluated using a consistent process and objective weighted average score. Contractors are then preferentially ranked based on the scoring and evaluation process. The highest weighted average scores are preferred. However, the highest score does not necessarily mean a contract award.

Contractors deemed HSE unacceptable are generally contacted and advised of the deficiencies that impacted their scoring and suitability. They are usually encouraged to upgrade their HSE information in the prequalification database for more favorable consideration in future SOWs.

To ensure evaluation consistency for larger SOWs, it may be necessary to provide contractors with a specific Contract HSE Plan Template (See Appendix 7—Sample Contract HSE Plan Template for SOW). Without this template contractors generally include their entire HSE manual in their bid, which typically includes a series of standards, safe work practices, check sheets and forms. This muddles the evaluation, requiring the owner's HSE resource to assume the contractor's ability to develop an appropriate SOW HSE Plan from the information provided.

The absence of a Contract HSE Plan Template impacts HSE evaluation and scoring as follows:

- Increased workloads for the owner's HSE team, constraining resources.
- Inaccurate evaluation of contractor HSE abilities and competencies.
    - Frequently, requisite HSE competency documents are purchased off the shelf without the contractor knowing the content or personnel within its workforce who can execute them.
- An inadequate comparison to contractors whose HSE Plan is available.
    - Comparing apples to oranges
- Eventual sub-optimal HSE performance during the SOW execution.

A Hazard Identification and Risk Mitigation Template may accompany the SOW HSE Plan Template to ensure the contractor's approach to identifying, assessing and mitigating risks is aligned with the owner's approach (see Appendix 8—Hazard Identification and Risk Mitigation Template for the SOW).

HSE Acceptable Contractor

HSE acceptable contractors are SOW candidates. Their selection depends on collective evaluation and scoring of various evaluation criteria. A structured evaluation may resemble a decision matrix as shown below, where Contractors 1 and 2 are scored similarly.

In such a case, either contractor may be acceptable to perform the SOW. This process is objective, consistent, and defensible.

Contractor selection will subsequently be highly influenced by the contractor's SOW cost estimate.

Evaluation Criteria	Weighted Score %	Contractor 1	Contractor 2	Contractor 3	Contractor 4
Engineering and Design	**25**	22	20	17	12
Quality – Reliability and Integrity	**25**	18	20	18	16
HSE/EHS	**30**	26	27	20	20
Commercial (Pricing)	**10**	8	8	6	6
Other Relevant Stakeholder Groups Criteria	**10**	9	7	7	7
Total	**100**	**83**	**82**	**68**	**61**

> **NOTE:** Weights for evaluation criteria may vary within the organization and across organizations.

Kick-Off Meeting—Provide Contractor with HSE Deliverable Checklist

Once a contractor has been selected for the SOW and has been awarded the contract, a kick-off meeting is coordinated by the procurement or supply chain management (SCM) organization with the selected contractor and owner representative stakeholders. The kick-off meeting is designed to communicate key SOW requirements, expectations, and deliverables to the contractor.

Organizations use this opportunity to reiterate the following:

- SOW technical deliverables
- SOW HSE requirements
- Known SOW hazards and risks
- Priorities for SOW execution—HSE, quality, and schedule

> **NOTE:** When HSE, Quality, and Schedule are adequately managed, budgets and cost control are maintained with fewer unplanned events, fewer rework requirements, and no idle time that cost- and schedule-burden the SOW delivery.

HSE leaders use this opportunity to provide the contractor with HSE deliverables specific to the SOW.

Owner Completes Pre-Mobilization Checklist Review with Contractor

The SOW owner typically reviews a pre-mobilization checklist with the contractor to ensure readiness for safe, reliable, and efficient SOW delivery (Appendix 9—Sample Pre-Mobilization Checklist). Owner and contractor project (SOW) leadership personnel systematically review requirements detailed in the pre-mobilization checklist and apply them to the SOW.

Where applicable, the contractor must provide the owner with verifiable evidence that requirements are met. The contractor must make compliance verification evidence available to the owner prior to mobilization based on risk exposures. Occasionally, at the discretion of the owner, unfulfilled requirements may be addressed by the contractor post-mobilization based on risk exposures and implemented risk mitigation plans.

For smaller, lower risk contracts and SOWs, the pre-mobilization checklist process and rigor may be significantly lighter.

HSE Approved Contractor (for Specific SOW)

Contractor HSE evaluation requires significant effort by the HSE organization to ensure an effective HSE process is implemented and executed. This can be updated in the prequalification database to streamline the HSE review process if the contractor is considered for additional or new work with the owner.

Care must be taken to ensure a full contractor HSE review is undertaken within an approved cycle—perhaps on a 2- to 3-year cycle if the contractor frequently performs work for the owner. This is commonly overlooked when the contractor performs well within the organization. However, the contractor may be servicing multiple clients (owners), and sub-optimal performance at another client's site may help indicate impending HSE performance challenges. HSE approval review exercises help identify and mitigate potential issues before they manifest.

HSE Alignment Workshop Conducted with Contractor as Applicable

An owner-led HSE Alignment Workshop is a high-value yet frequently underutilized process for HSE performance improvements with contracting workforces. Prior to mobilization to the worksite, the owner conducts a 2–4 hour Alignment Workshop with key contractor stakeholders to ensure the following:

- All contractor employees are aware of the owner's HSE requirements and expectations during work execution. For example, the owner may reinforce that safety comes before anything else.
- Contact points between contractor and owner personnel, where safety is concerned, are reaffirmed.
- HSE events reporting and investigation requirements are known.
- Hazards, risks, and mitigation actions are known.
- The contractor is aware of execution priorities—HSE, Quality and Schedule, in that order.

This alignment workshop also communicates the owner's emphasis on safety, wellbeing and the desire to ensure excellence in the execution of work.

Site-Specific Orientation and Training Completed by Contractor

Many contractors are not aware of the hazards and risks associated with the owner's worksites. Furthermore, there may be site-specific training that the contractor must receive to perform work safely and efficiently.

Orientation training may be awareness type training, whereas site-specific training may address certain types of hazardous work (e.g., confined space entry work) and may require competency verification and assurance.

Site-specific training may differ from orientation training.

Mobilized Contractor

A mobilized contractor is ready to move to the worksite. The contractor has been evaluated and offered a contract, undergone pre-mobilization reviews, closed gaps identified in the pre-mobilization reviews, received orientation and site-specific training and is ready to begin executing the work.

It's important to note that some SOWs present significant risks and hazards during mobilization. For example, drilling rig movement to a new owner's drilling site presents logistical challenges such as:

- Road traffic management
- Hazards **A**bove (e.g., power lines), **B**ehind (e.g., blind spot visual obstructions), **B**elow (e.g., soft roads), and **I**nside (e.g., poor ventilation)
    - It's imperative to address all **ABBI** hazards and concerns
- Logistics associated with moving heavy equipment
- New and/or inexperienced workers
- Rural environments

Additional supervision helps ensure an event-free mobilization to the new worksite.

30-Day Verification of HSE Plan Implementation

The owner should execute a 30-day post-mobilization verification of HSE Plan implementation for the SOW. This verification exercise is a critical component of the process for the following reasons:

- Gaps in the process are identified and corrected early
- Contractors can better understand and meet the owner's expectations and requirements
- Normalization of unsafe practices is mitigated

Refer to Appendix 10 for 30-Day Contract HSE Implementation Assessment.

> **NOTE:** Occasionally, security and/or quality may be included in the HSE Plan that may be referred to as the HSSE/HSSEQ Plan.

Contractor Performance Management

Managing contractor performance is perhaps the biggest opportunity for improving HSE performance. High-value performance improvement opportunities include the following:

- Improving leadership visibility
- Treating all workers as equals where safety is concerned
- Communicating performance evaluation criteria early
- Communicating priorities—safety, quality, and schedule, in that order
- Establishing and stewarding KPIs—focus on leading indicators
- Initiating contractor safety audits
- Initiating regular contractor safety forums.
- Relationship management—applying the ABCD of trust
- Celebrating small wins and milestone achievements
- Sharing learnings

These requirements are further detailed in Chapter 4.

Contractor Performance Evaluation—Prequalification Database Updated

Upon completion of the SOW deliverables, the owner must conduct a performance evaluation of the contractor before demobilization. It is significantly more beneficial for this to be a joint owner-contractor interactive discussion prior to the contractor leaving the owner's worksite.

An interactive discussion helps in achieving the following:

- Clarification of owner and contractor performance perspective differences
- Knowledge sharing and capture
  - Agreeing on what worked well and what did not
  - Documenting potential best practices for similar or repeated work
- Identification and agreement on improvement opportunities by both the owner and contractor.

Performing this exercise after the contractor is demobilized offers little value, often reflecting the owner's perspective rather than a joint perspective.

---

4 CONTRACTOR PERFORMANCE MANAGEMENT

Performance Management Requirements for All Contractors

Selecting the Right Contractors

A robust contractor selection and mobilization process includes the following:

- Prequalification
  - Using contractors with demonstrated commitment to safety
  - Obtaining information on safety programs and prior performance evaluated for satisfactory HSE performance
- Contract preparation and award
  (HSE participation and involvement essentials)
  - SOW definition
  - Clearly defined HSE requirements
  - Provisions for corrective and mitigative actions for contractor failure to meet the owner's HSE requirements
- Pre-mobilization meeting and contract kick-off
  (See Appendix 9 for sample Pre-Mobilization Checklist.)
  - HSE Alignment Sessions
  - Contractor orientation and training verification
    - General worksite information
      - Rules
      - Procedures
    - Work area-specific training (where applicable)
  - Maintained site-specific training records of contractor representatives
  - Equipment status (licenses, maintenance records, load capabilities, etc.)
  - Contractor safe work practices and procedures

> **NOTE:** Owners and contractor SOW leaders should pay particular attention to mobilization (excitement) and demobilization (what's next) concerns.

Contractor HSE Performance Management

Managing contractor performance helps owners achieve on-budget, on-schedule delivery of SOW deliverables. Careful attention to planning and execution management leads to safe, reliable, and efficient performance. Among the high-value performance management measures are the following:

- Improving leadership visibility
  - Regular planned visits to facilities and frontline with predetermined frequency
  - Frequent safety visits and inspections (including contractor leadership)

DOI: 10.1201/9781003728467-4

  - Owner visits to contractor shops and facilities
  - More two-way communication where safety is concerned
    - Ensure frontline workers are engaged
    - *Listen* to the concerns of frontline workers
  - Demonstrate and promote a *no blame* culture (i.e., remove any perception of punitive implications for reporting incidents).
  - *Listening Moments*—conduct active listening sessions/ lunches.
- Defining and communicating performance evaluation criteria early
- Communicating priorities—safety, quality and schedule (in that order)
- Establishing and stewarding KPIs—focus on leading indicators
- Initiating contractor safety audits
- Treating all workers as equals (where safety is concerned)
- Initiating regular contractor safety standdowns and forums
- Managing relationships with contractors and applying the **ABCD** of trust
- Celebrating small wins and milestone achievements
- Sharing learnings among the entire workforces

These performance management processes are detailed further in this chapter.

Leadership Visibility

Among the most powerful contributors to improving contractor safety performance is leadership visibility at the worksite and the frontline. Leadership visibility demonstrates to the workforce that the owner cares about the safety of the workforce. Visibility is more powerful when done jointly with the contractor's leaders.

Leadership visibility must be planful and purposeful and require strategic intent as detailed below.

Visibility Activity	Details
Scheduled Frontline Visit	• Frontline visits are a powerful leading indicator for improving HSE performance • Should be planned and conducted with contractor counterpart • Must be structured—See Appendix 11 for Six Steps for an Effective Frontline Visit
Frequent Safety Visits and Inspections	• Multiple types of safety inspections are available for participation by leaders that should be supported by a documented response process. ○ Often, an inspection form linked to a corrective actions management process

	• Among the field inspections with accompanying forms may be the following: ○ Aerial Work Platform (AWP) Pre-Use Inspection Checklist and Operator Qualification ○ Compressed Gas Cylinder Inspection Checklist ○ Daily Excavation and Trench Inspection Report ○ Daily Maintenance of Mobile Equipment ○ Earthmoving Equipment Inspection Checklist ○ Equipment Inspection ○ Eyewash Station Maintenance ○ Fire Extinguisher Maintenance ○ Formal Site Inspection and Report ○ Industrial Forklift Inspection Checklist ○ Monthly Office and Facilities HSE Inspection Checklist ○ New Employee Orientation Checklist ○ Quarterly Oxygen/Fuel Equipment Inspection Checklist ○ Vehicle Inspection ○ Scaffold Inspection Checklist ○ Vendor Truck Loading and Unloading Checklist
Visits to Contractor Shops	• Visits to the contractor's shops and premises provide valuable insights into the HSE culture of the contractor. ○ Poor housekeeping is an immediate concern • Informal chats with contractor employees tend to generate valuable insights on what the owner should focus on during execution of the SOW.
Worker Communication and Engagement	• Two-way communication is fundamental to effective frontline communication. • Listening to frontline workers tells us a lot about where potential high-risk exposures may exist. • Talking about things that are important to the worker where safety is concerned.
Promote a *No Blame* Culture	• Creating a culture where everyone is comfortable to proactively raise HSE concerns where they can be addressed before events occur. • Eliminating blame—focusing on fixing problems before events occur. • Avoiding punitive consequences.

Active Listening and Listening Moments	• Listen—don't just wait for your turn to speak.
	• Implement listening moments—eliminate the pen and listen to the workforce. Workers tend to become cautious when a leader documents their concerns.
	• Seek informal moments like a lunchroom meeting with owner-sponsored lunches to discuss HSE concerns with the workforce.
	  ○ Ensure HSE concerns are informally recorded, prioritized, and addressed.
	    ▪ Ultimate success comes from acting on the issues raised by the workforce.
	    ▪ You may not be able to act on all concerns, but prioritization and action generate trust and build credibility.

Define and Communicate Performance Evaluation Criteria

Contractor performance (including HSE) is best managed when the evaluation criteria are communicated to the contractor at the start of work. See Appendix 12—Sample List of Common Performance Evaluation Criteria applied in contractor performance management. Communication of applicable evaluation criteria early in the execution of the SOW ensures the contractor is aware of the requirements and expectations important to the owner in the execution of the SOW.

Communicate Priorities—Safety, Quality, and Schedule

Contractor safety management is the result of clear and effective communication of the owner's priorities during SOW execution. Owners should communicate priorities of safety, quality, and schedule (in that order). Details of this prioritization are provided below.

Priority	Details
Safety	• Safe operations result in a motivated and highly productive workforce. • Direct and indirect costs of safety events can be high. • Unsafe work environments adversely impact the image of the organization. • Unsafe work environments have difficulties attracting the best and brightest of the workforce. • Often, safe work performance results in on-budget, on-schedule delivery of work.
Quality	• Quality work eliminates the need for costly repairs and rework. ○ Poor quality work often results in costly rework, delaying schedules. • Quality concerns lead to longer-term problems in asset reliability, integrity, and potential loss of primary containment (LOPCs).

Schedule	• Schedule delays often lead to major cost escalations. For example, an idle crane can cost an owner tens of thousands of dollars per day of delay. • Careful planning and execution of concurrent work can avoid events and cost escalations. • A safety event has the potential to delay schedules when productivity is impacted—for major events such as serious injuries and fatalities (SIFs) and fires and explosions from LOPCs. • Delaying hazardous conditions in the workplace can reduce the impact on schedules. ○ For example, delaying permanent power and process fluids (hydrocarbons and natural gas) have positive impacts on worker productivity.
Cost (Budget)	• Generally, when safety, quality, and schedule are well-managed, costs often align with budgets and require little attention from leaders in cost overrun management and accommodations.

Conduct 30-Day HSSE Plan	It is important that the implementation of the SOW HSE plan is assessed for contractor compliance early in the SOW execution. This process helps to identify gaps in compliance and allows the implementation of corrective actions early enough to avoid unintended consequences. See Appendix 10—30-Day HSE Plan Implementation Assessment for proactive interventions and correction of deficiencies.

Establish and Steward KPIs—Focus on Leading Indicators

Establish and Steward KPIs—Focus on Leading Indicators

Establishing and communicating Key Performance Indicators (KPIs) and measures against which the contractor's performance will be measured is essential for effective stewardship of the contractor's performance. See Appendix 13—Sample Contractor Safety Management KPIs.

Performance measures and KPIs should conform to the following:

- Be communicated to the contractor prior to the start of work
- Be agreed to by both parties
- Focus on the following:
  - Proactive leading indicators which are ***actively managed***
  - Reactive lagging indicators that are ***actively monitored***
- Be limited to selected high-value indicators (leading and lagging)
- Conform to the SMARTER Principles (**S**pecific, **M**easurable, **A**chievable, **R**ealistic, **T**ime-Bounded, **E**xplainable, and **R**elative).
- Be supported by effective methods for the following:
  - Data collection methods and sources — and who collects the data.

    - Methods for calculation—applicable formula.
    - Frequency of analyses and communication, including formats for communication—trends, pie charts, data labels, etc.
    - Unacceptable zones and performance levels with early intervention zones.
- KPIs setting should follow the PDCA Principles as shown in Figure 4.1.

Figure 4.1: Linking KPIs setting to the PDCA model.
Source: OEMS Erudite Corp (2025).

SMARTER Attributes of KPIs are demonstrated in Figure 4.2.

	Attribute	Example	KPIs Stewardship Guidance
S	Specific	85% of mechanics are competency assured.	1. Data for measuring KPIs should be simple and easy to collect. 2. Data should be able to produce decision-making knowledge. 3. Tools and processes for collecting and analyzing data must be accessible. 4. Knowledge generated must be accurate, relevant and timely.
M	Measurable	Is 85% a measurable target?	
A	Achievable	Can we achieve 85%?	
R	Realistic	Is it realistic to have 85% of mechanics competency assured?	
T	Time-bounded	Over what period do we wish to achieve 85%?	
E	Explainable	Can we explain why 85% is important?	
R	Relative	KPIs can transition from absolute numbers to growth percentages.	

Figure 4.2: Attributes of SMARTER KPIs.
Source: OEMS Erudite Corp (2025).

Figure 4.3 demonstrates the relationships between leading and lagging measures and the consequences associated with the various measures.

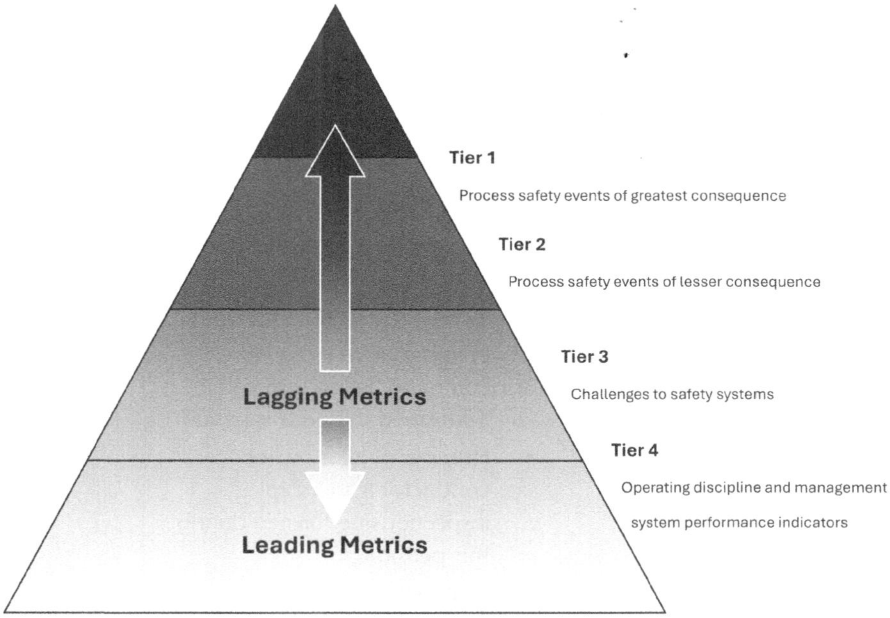

Figure 4.3: Transition from leading to lagging process safety metrics.
Source: Adapted for grayscale visibility from CCPS (2022).

Leading and Lagging Indicators—Attributes

Leading and lagging indicators are extremely important for improving contractor HSE performance. Careful attention must be placed on selecting the right indicator and ensuring the contractor is not overly taxed (tasked) in collecting and analyzing data, rather than focusing on safe, reliable, and efficient SOW delivery. The table below summarizes the approach and value of leading and lagging KPIs.

Leading Indicators	Lagging Indicators
Proactive focus—seeing what's ahead.	**After the fact**—rearview mirror looks.
Before an incident occurs—provides opportunities for proactive interventions and course corrections.	**The incident has occurred**—too late to prevent.
Learning from events—learning from external sources to prevent *occurrence*. Translating learning to action at the frontline to prevent events.	**Shared learning**—now an external source to other owners and contractors designed to prevent *recurrence*.
Examples: • Audit findings closed versus identified • % of permits properly filled out • % of overdue preventive maintenance (PM) orders • % of standard operating procedures (SOPs) and safe work practices (SWPs) updated and approved	Examples: • Injury numbers and frequencies ○ Recordable injury frequency (RIF) ○ Total injury frequency (TIF) ○ Lost time injury frequency (LTIF) • Spills frequency • Motor vehicle incidents (MVIs) • Fatalities

High-Value Leading and Lagging Contractor Safety Indicators

KPIs and performance indicators must be high-value and easy to collect and analyze. High-value KPIs are those that minimize risk and consequences to the organization. KPIs with the highest safety improvement impacts are defined in this section.

Leading Indicators	Lagging Indicators
• Contractor HSE audit findings risk-prioritized with corrective actions closed versus identified • Number of joint owner-contractor leadership visits performed during this period • Number of joint owner-contractor leadership inspections performed during this period • Number of near-misses and incidents recorded, investigated and with assigned corrective actions • Number of HSE communication sessions and listening moments performed during the period	• Contractor injury numbers and frequencies (i.e., RIF, TIF, LTIF) • Spills frequency • Motor vehicle incidents (MVIs) • Serious injuries and fatalities (SIFs) • Fatalities

Other Recommended High-Value Leading Indicators

Other high-value leading indicators include the following:

Leading Indicators	
Safety Training and Certifications Completion Rates	**Description**—percentage of contractors who have completed mandatory safety training and hold relevant safety certifications.
	Importance—ensures contractors are prepared and aware of safety protocols before beginning work, reducing the likelihood of accidents.
Pre-Job Safety Planning (JSA/Task Hazard Analysis)	**Description**—percentage of jobs with completed job safety analysis (JSA) or task hazard analysis (THA) before starting work.
	Importance—identifies potential hazards before work begins can help eliminate risks and ensure proper controls are in place.
Number of Non-Routine Tasks and Work Activities Performed with Procedures and Safe Work Practice In-Hand	**Description**—percentage of tasks and work activities with procedure and SWP reviewed by the work group prior to work execution.
	Importance—reviewing procedures and SWPs prior to executing work helps drive operational discipline and excellence.

Safety Meetings and Toolbox Talks with Leadership Attendance	**Description**—percentage of contractors attending regular safety meetings or toolbox talks.
	Importance—reinforces safety culture and keeps safety at the forefront throughout the project, enhancing communication of potential hazards.
Contractor Safety Performance Metrics Management and Communication	**Description**—tracking of key performance metrics, such as the number of near misses, safety observations, and unsafe acts reported by contractors.
	Importance—proactive reporting of unsafe conditions or behaviors indicates a robust safety culture and helps mitigate risks early.
Personal Protective Equipment (PPE) Compliance	**Description**—percentage of contractors observed wearing appropriate PPE on the job site.
	Importance—ensures workers are using the proper protective gear, reducing the risk of injury from common hazards (e.g., falling objects, electrical hazards).
Safety Behavior Observations	**Description**—number of positive and corrective safety behavior observations made by supervisors or safety officers.
	Importance—observing and reinforcing safe behavior on-site helps prevent accidents and encourages a proactive safety culture.
Contractor Safety Orientation Completion	**Description**—percentage of contractors who have completed the company's safety orientation before beginning work.
	Importance—ensures that all contractors understand the company's safety expectations and specific job site hazards.
Compliance with Safety Regulations and Best Practices	**Description**—monitoring the contractors' adherence to safety regulations, codes, and best practices set by regulatory bodies (e.g., OSHA).
	Importance—compliance is essential to ensure contractors meet industry safety standards and to reduce legal and safety risks.

Initiate Contractor HSE Audits (Assessments)

Initiate Contractor HSE Audits

Collaborative contractor HSE auditing is a powerful process for generating significant improvements in contractor safety performance. The owner and contractor work collaboratively to proactively identify gaps in compliance and close them before an event occurs.

Contractors are generally excited to identify and address improvement opportunities identified in the auditing process. Figure 4.4 demonstrates the value of collaborative contractor auditing and the performance improvements generated.

In this collaborative auditing process, the following approach was adopted:

- Joint contractor and owner involvement in the audit process.
- Contractors were forewarned and engaged.
- Punitive consequences were eliminated from the process—focus on identifying at-risk gaps and closing them before events occurred.
    - Gaps identified were risk-assessed and prioritized for closure.
- An open and transparent approach to auditing—contractors encouraged to be open without fear of consequences.
- Known areas for improvements were highlighted, focused on, and addressed.

As shown below, when contractors are audited collaboratively, the HSE performance improvement can occur rapidly with potentially excellent performance. As shown in Figure 4.4, within six months (June 08–Dec 08) of the introduction of collaborative auditing, strong HSE performance improvement was experienced.

> **NOTE:** Conventional (invasive/policing) contractor auditing promotes evasion and hiding of known gaps in compliance. Consequences for non-compliance are generally punitive, and events are more likely to occur with invasive auditing.

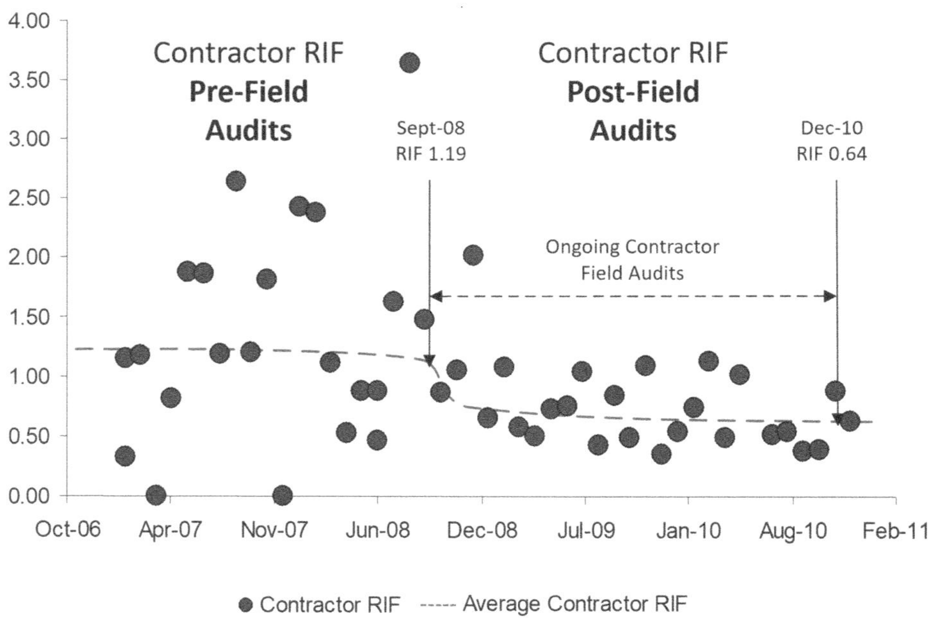

Figure 4.4: An almost 100% improvement in HSE performance from collaborative auditing. Source: Safety Erudite Inc. (2011).

Conventional vs. Collaborative Contractor Safety Management Audits

Attributes of conventional and collaborative contractor audits are provided below.

Conventional Audits	Collaborative Audits
• Appears to be a policing exercise. • Auditor appears to be poking holes in compliance. • Contractors seek to hide known gaps and pursue evasion tactics to avoid punitive consequences. • Contractors may face punitive consequences for gaps identified. • Contractors may not be forewarned of audits by owners, causing surprise and unpreparedness for an audit. • Contractors may become defensive, and owner-contractor relationships may become strained because of punitive consequences.	• A proactive and shared interest in improving performance. • Auditors test the robustness of the contractor's compliance measures. • Joint identification of improvement opportunities. • Encourages open/frank discussions. • Contractor encouraged to surface issues and gaps in compliance to close the gaps before events occur. • Focus on closing gaps versus punitive consequences. • Collaboration and trust relationship building between the contractor and owner. • Audits are planned, with areas of concern identified and opportunities for evidence gathering and compliance verification afforded to the contractor.

Collaborative Auditing Questioning Approaches for Contractors

The following are recommended questioning approaches after putting contractors at ease:

- Open-ended questions that encourage dialogue and explanations.
- No trick questions.
- Avoid saying or using "**No**." This shuts people up.
- Avoid interrupting the interviewee during responses—let them finish.
- Frame questions to understand the contractor's operations and procedures for collaborative review.
- Adopt a trust and verify approach to evidence collection and verifications.

Audit Observations and Findings

The goal of an audit is to proactively identify gaps in HSE compliance and implementation of improvement opportunities. Consequently, auditors (assessors) seek observations and findings that demonstrate the following characteristics:

Characteristic	Details
Relevance	Observations must be relevant to HSE plan requirements/ deliverables and do not reflect aspects relevant to other non-HSE requirements.
Materiality	Material observations are those that may impact performance or precipitate safety events. Material observations are linked to **PACA** as follows: • **P**rocedures—availability/accessibility/accuracy/use. • **A**ccountability—compliance gaps or deficiencies may result from unclear requirement accountability assignments. • **C**ompetency—training and competency assurance not completed or confirmed. • **A**ssurance and verification—no one checking to verify requirements are being met.
Credibility	Observations, when defined, should be credible. They must be ***factual and free of the assessor's opinions***. Hearsay and speculations should not be entertained in the assessment process.
Objectivity	Observations must be objective and quantifiable. Subjectivity (assessor's views) in observations should be discouraged.

Documenting Observations and Findings

The auditor should ensure the following components are present in the documentation process, as findings may require corrective actions, and accurate interpretations will be required.

What to Document	How to Write it
• **Relevant information**—document findings and observations information relevant to the requirements defined in the SOW. • **Material**—observations/findings should reflect a material impact on safety, reliability, and efficient performance of the business • **Credible**—factual information—not opinions (what you see and hear from the interviewee). • **Objective**—measurable and relatable—does not rely on perception. • **Corrective Actions/ Recommendations**—if the auditor is an appropriate SME, provided corrective actions and recommendations to close gaps identified is acceptable. Alternatively, this should be assigned to an SME.	• **Use simple language**—simple and common words. • **Be specific**—identify specific details of compliance/ non-compliance. • **Be clear**—write clearly–remove ambiguity in sentences. • **Support**—with drawings/images where possible. • **Avoid recommendations and solutions** (except for sharing best practices). • **No Opinions**—document only the facts.

Typical Collaborative Auditing Process

A typical collaborative audit process is provided in Figure 4.5. Collaborative audits generally lead to strong contractor-owner relationships with a joint interest in executing work safely, reliably, and efficiently.

- In the pre-audit phase, contractors should be notified well in advance. The contractor prepares for the audit, gathers evidence for compliance verification and is generally ready for the audit.
- In the audit phase, contractors are placed at ease and encouraged to share information and provide evidence to support compliance with HSSE requirements defined by the owner. The contractor may raise areas of weakness in compliance.
- In the post-audit phase, corrective actions are risk-prioritized, assigned, and stewarded to completion.

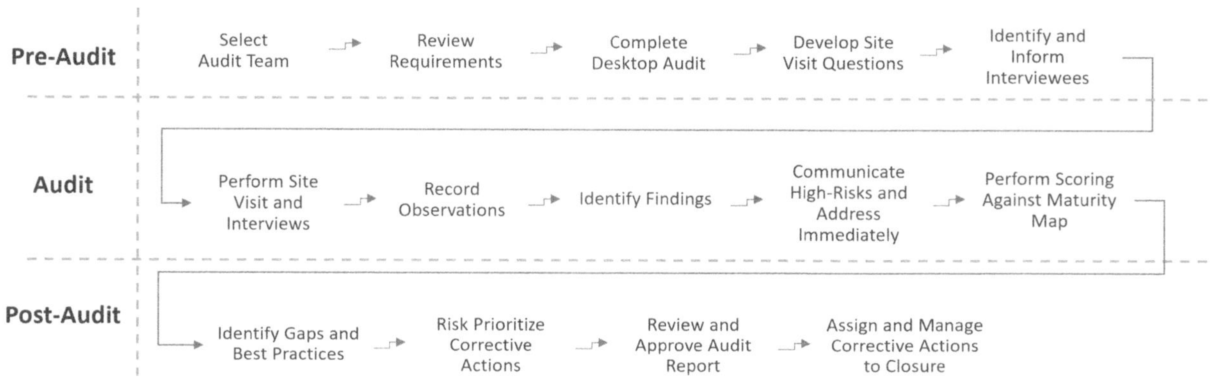

Figure 4.5: Steps in the collaborative auditing process.
Source: OEMS Erudite Corp. (2025).

Treat all Workers as Equals (Where Safety is Concerned)

Treat all Workers as Equals

Where safety is concerned, there should be absolutely no difference between how we treat employees and contractors. Contractors and employees alike work to survive, and it's our responsibility as leaders to ensure their safety.

Consequently, leaders should ensure the following:

- All workers are aware of their right to refuse unsafe work.
- All workers are encouraged to intervene in unsafe work or conditions with no punitive consequences.
- PPE, the last line of defense, is available to all workers.
- A governing principle: *no work is so urgent or important that leaders cannot take the time to ensure it is safe to perform.*
- Safety is always prioritized ahead of production.
- Only trained and competent workers are allowed to perform work independently.
- Inexperienced workers are supervised and mentored during work.
- Communication is conducted in the language of the worker.
    - Apply *showing* instead of *telling* in the communication process.

Initiate Regular Contractor Safety Stand-Downs and Forums

Initiate Regular Contractor Safety Stand-Downs and Forums

Contractor safety stand-downs and forums are high value activities in ensuring safe, reliable, and efficient work. Essentially, safety stand-downs pause work to focus on improving safety. Typically, however, both these measures are underutilized by owners. Owners and contractors should collaborate to ensure both these activities are more frequently used in the workplace—and used proactively rather than reactively. Details about safety stand-downs and forums are as follows:

Safety Stand-Downs	Work is paused to focus collective attention on safety. Common attributes of effective safety stand-downs are as follows: • Planned—the most effective use of safety stand-downs is in a planned, proactive fashion. Utilizing safety stand-downs only after events have occurred creates a negative image. • During work stoppage—the entire workforce should be able to participate, undistracted. This reinforces the sentiment that safety is prioritized. • Leadership presence—visible leadership participation models the importance of safety at all organizational levels. • Focused—specific safety topics, hazards, or incidents should be highlighted to allow maximal worker absorption of information. • Interactive and engaging—as discussed, two-way communication fosters active contractor participation and value maximization. • Clear messaging—a simple, direct message is more likely to stick. Visual aids, demonstrations, and even storytelling help appeal to various learning types within the workforce. • Organized—there should be documented attendance, formalized feedback collection, and possibly updated policies distributed. • Reinforces safety culture—an overarching theme of a robust safety culture should be prominent. It's okay to highlight safety successes as well for positive reinforcement.
Contractor Forums	Contractor forums are typically structured meetings between owners and contractors to align safety, expectations, and performance. Common effective contractor forum attributes include the following: • Regular and structured—there should be a clear cadence of forums (e.g., quarterly, semiannually) with a clear, predetermined agenda. • Two-way communication—contractors and owners should exchange ideas and concerns to facilitate mutual engagement and benefit.

	• Safety-focused—an emphasis on safety topics, lessons learned, and reinforcing safety culture should be apparent. • Benchmarking—when considering contractor performance, industry standards may be used as a reference. • Relationship building—collaborative problem-solving and open communication should strengthen the owner-contractor relationship. • Training and education—owners can utilize these opportunities to help keep contactors up to date on industry standards, technologies, and procedures. • Recognition—positive contractor behaviors should be acknowledged and even rewarded where applicable. • Shared learning—best practices, lessons learned, and incident investigation case studies are all high-value inclusions. • Goal alignment—contractors should be encouraged to take ownership of their safety performance, aligning with the owner's broader standards.

Improvement Opportunities for Safety Stand-Downs and Forums

Applying both safety stand-downs and forums can more greatly contribute to the delivery of safe, reliable, and efficient business performance should they be used as follows:

	Guidance

Stand-Downs	• Highlight good practices and identify improvement opportunities.
	• Focus on genuine care for the entire workforce.
	• Remove punitive consequences, instead focusing on collaborative performance improvements.
	• Provide an opportunity for workers and contractors to pause, focusing specifically on safety/HSE issues.
	• Foster an open environment for workers to share safety concerns, near-miss incidents, and suggestions for practice improvement.
	• Reinforce the importance of safety as a top priority and value.
	• Allow contractors to highlight specific or emerging hazards that may not have been fully addressed in routine safety training.
	• Demonstrate the following:
	    ○ Visible leadership commitment to safety
	    ○ Leadership participation, showing workers they are valued and actively involved in improving safety performance—helping to enhance worker buy-in and engagement

Forums	• Create routine events for long-term contractors to share knowledge and best practices. • Allow contractors to be the architects of conversation. • Invite third-party or independent subject matter experts to address/resolve common concerns. • Provide an opportunity to reward performance excellence. • Encourage greater emphasis on HSE.

Manage Relationships with Contractors—Apply the ABCD of Trust

Manage Relationships with Contractors—Apply the ABCD of Trust

Credibility and trust are important to any relationship, and they are extremely important in contractor-owner relationships for performance excellence. Trust and credibility come from consistent effort and can be eroded with a single act of indiscretion.

Where the contractor-owner relationship is concerned, trust and credibility come from doing the *right thing*. Sometimes, the right thing may be different from what is agreed upon and requires strong leadership to make the right decision. However, when done well, relationships are strengthened, and performance becomes stronger and repeatable.

The ***ABCD of trust***, as detailed below, helps create and sustain trust, credibility, and positive relationships between contractors and owners.

ABCD	Leadership Behavior	What We Mean
A	**ABLE**—demonstrate competence.	• Produce results • Make things happen • Be competent and capable • Know the organization and set people up for success
B	**BELIEVABLE**—act with integrity and be credible.	• Be honest in dealing with people; be fair, equitable, consistent, and respectful • Value-driven behaviors reassure employees that they can rely on their leaders • Act with Integrity
C	**CONNECTED**—demonstrate genuine care and empathy for people.	• Understand and act on worker needs; listen and share information. Be a real person • When leaders share a little bit about themselves, it makes them approachable • Show genuine care for others • Be responsive to the needs of others
D	**DEPENDABLE**—follow through on commitments.	• Say what you will do and do what you say you will • Being organized reassures followers • Be reliable

Source: Adapted from Ken Blanchard ABCD of Trust Model (2025) and Olmstead, C. (2024).

Building Credibility and Trust

As a leader, you have the power of choice: build trust or create mistrust. In CSM, trust is either built or undermined through your actions and behaviors. The guidance provided below delineates methods of directly influencing trust—either positively or negatively—within the workforce.

Trust	Guidance
Building Trust	• Giving credit to those who deserve it • Actively listening • Setting clear and achievable goals • Being honest • Following through on commitments • Demonstrating genuine care for your people
Creating Mistrust	• Limiting communication • Being dishonest • Breaking confidentiality • Taking credit for others' work • Assigning blame without understanding root causes

Celebrating Small Wins and Milestone Achievements

Celebrate Small Wins and Milestone Achievements

Celebrating small wins and milestone achievements with the contractor workforce showcases performance and highlights the shared success of contractor and owner collaboration. Recognizing achievements positively impacts contractor safety performance in the following ways:

- Recognizing small wins helps create a positive work environment.
  - When contractors feel appreciated for their HSE efforts, they are more likely to stay motivated and committed to maintaining safety protocols.
  - This increased morale leads to a heightened sense of responsibility and a more proactive approach to HSE.
- HSE importance is reinforced by making it a part of the owner's culture.
- Tangibly recognizing achievements, such as reaching a target number of accident-free days, encourages contractors to continue following safe practices.
- Acknowledging individual or team achievements can foster friendly competition among contractors, encouraging them to strive for the next milestone.
  - This can lead to a continuous improvement cycle, where contractors push each other to stay safe and achieve higher standards.
- When HSE achievements are celebrated, it shows that the organization values safety as much as other business goals.
  - Over time, this helps build a strong safety culture where safety becomes a shared priority across the workforce, leading to better overall safety performance.

- Celebrating small wins gives contractors the confidence to improve and innovate their safety practices.
    - As each milestone is met, it demonstrates progress and encourages ongoing improvements to safety protocols, ultimately contributing to a safer workplace.

Share Learnings Among the Entire Workforce

Share Learnings Among the Entire Workforce

Sharing learnings across the organization and among the entire workforce creates an effective process for systematically eliminating repeat or similar events. Shared learning, however, must be organized and designed to influence human behaviors in a positive way.

Shared learning should retain the following attributes:

- Be simple and easy to follow
- Be accurate
- Focus on the following, where events are concerned:
    - A summary of what happened
    - A compilation of root and contributing causes
    - Corrective actions to be implemented to prevent a repeat event
    - Demographic information so that users can contact the initiator for additional information as required
- Be easily accessible via push and pull strategies to those requiring it
- Include assigned corrective and preventive actions to both contractors and owners
- Allow for owner, industry, and cross-industry sharing of knowledge and learning from both events and best practices
- Follow an organized process to ensure content does not introduce liabilities

Refer to Appendix 14—Sample Template for Sharing Learnings and Knowledge.

Value of Shared Learning to Contractor Safety Performance

Shared learning helps improve contractor safety performance in the following ways:

- Shared learning allows contractors from different teams or projects to exchange safety experiences and best practices.
    - When contractors learn from each other's successes and challenges, they can adopt strategies that have proven effective, reducing the likelihood of accidents or unsafe practices on other projects.
- Encouraging shared learning creates an environment where safety is seen as a team effort rather than an individual responsibility.
    - This collaborative approach strengthens the overall safety culture, as contractors are more likely to support one another in maintaining and improving safety standards.

- By sharing HSE insights and lessons learned from past incidents or near-misses, contractors can identify potential risks that they may not have previously considered.
    - This proactive approach helps to mitigate hazards before they lead to accidents, ensuring a safer work environment for everyone involved.
- Shared learning promotes creative thinking and problem solving when it comes to safety challenges.
    - Contractors can discuss what strategies or tools have worked well in different situations, leading to innovative solutions that may not have been considered otherwise.
    - This helps to continually improve safety practices and adapt to changing circumstances.
- When contractors share their safety experiences, it encourages accountability within the team.
    - Knowing that their actions and decisions can be shared with others motivates contractors to adhere to safety protocols more closely and take ownership of their role in creating a safe work environment.

Manage Risks Associated with the Scope of Work

Overview

Perhaps the biggest opportunity in improving contractor safety performance is in the owner's approach to risk management and the demands placed on contractors regarding risk management. Multiple tools and processes are available to identify and mitigate risks. Owners should implement these processes effectively to ensure the workforce is protected during work execution.

Findings of a study performed and presented at conferences by IHS on root causes of events showed that approximately 28% of events across a large cross-section of organizations were directly related to ineffective hazard identification and risk management processes within the organization, as shown in Figure 4.6. Furthermore, other leading categories of causes of events were in themselves linked to hazard identification and risk management processes, as shown in Figure 4.6.

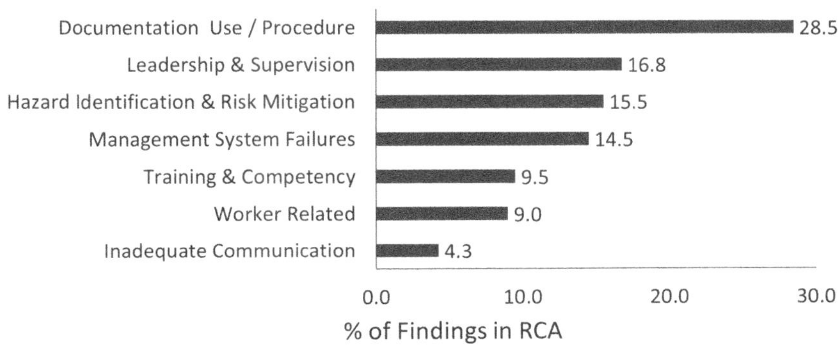

Figure 4.6: Summary root and contributing causes to major events in organizations. Source: OEMS Erudite Corp. (2025). Adapted from IHS (2014) Unpublished.

Underlying contributors to the categories of causes of events presented in Figure 4.6 are provided in Figure 4.7.

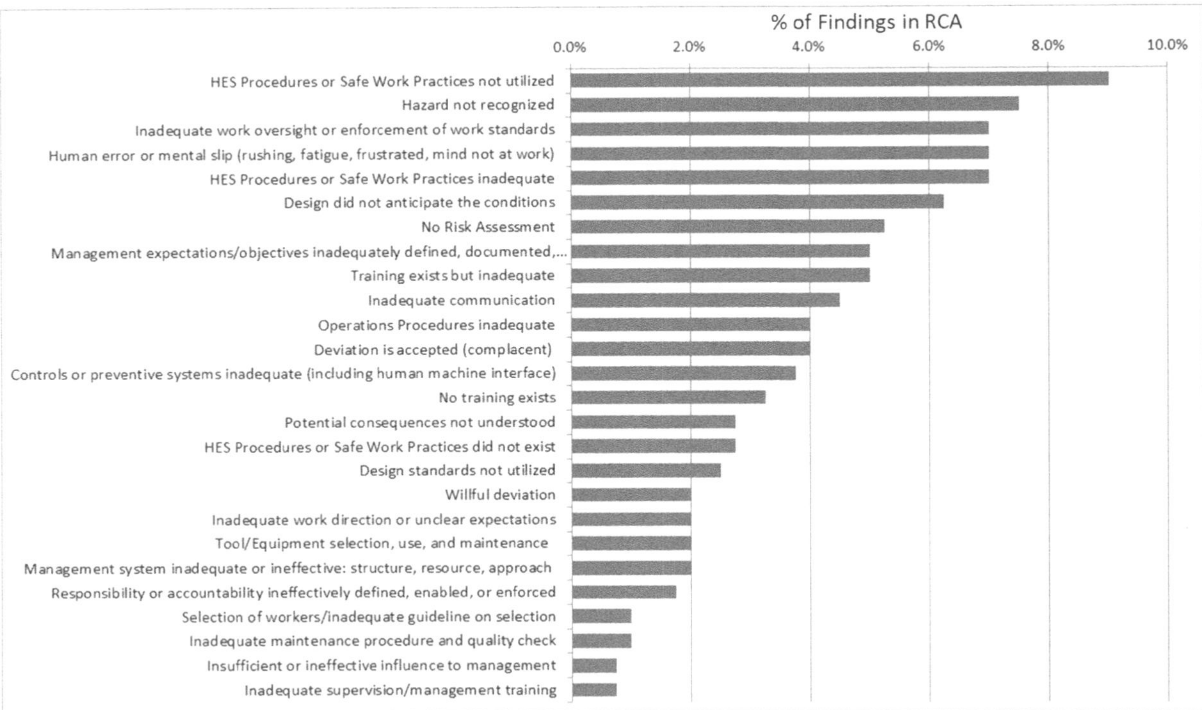

Figure 4.7: Underlying root causes to major events in organizations, as summarized in Figure 4.6.
Source: OEMS Erudite Corp. 2025. Adapted from IHS (2014) Unpublished.

Commonly Employed Hazard Identification and Risk Management Processes

Commonly employed hazard identification and risk management tools and processes employed by owners are as follows:

- Job safety analysis
- Task hazard analysis
- Field level hazard assessment
- Field level risk assessment
- A safe work permit (SWP) to work process
- Toolbox talks
- Safety alerts
- Risk matrices
- Risk registers

In all cases, during the execution of any work, known hazards and risks shall be communicated to the workforce (employees and contractors alike). Furthermore, contractors must be encouraged to do the following:

- Refuse all unsafe work (a ***right*** in some regions of the world)
- Intervene when unsafe conditions exist or unsafe acts are being committed
- Report unsafe acts or conditions without fear of punitive consequences

Communicating known hazards and risks associated with the contracted scope of work is best achieved using tools and processes as demonstrated in Appendix 8—Hazard Identification and Risk Mitigation.

Where contractors do not have a risk matrix, owners are encouraged to include their risk matrix with the bid documents so that all risks are assessed consistently for the SOW.

Appendix 15—Sample Risk Evaluation Guide (in the absence of an owner's risk matrix) for enabling effective risk assessment by the contractor.

5 CHALLENGES FACED BY CONTRACTORS AND OWNERS

Challenges Faced by Contractors

Contractor Challenges—Overview

Contractors face many common challenges, including the following:

- Competition–which leads to cost and profit squeeze
- Owner pressure for cheaper, better, and faster work–which leads to cost and profit squeeze.
- A transient workforce, always changing with new SOWs and work locations.
- Loss of skilled workers who are often pilfered by owner companies when the SOW is repeated by the owner.
- Scope changes and scope creep
    - Changing owner expectations and requirements
- Ineffective communication—methods, frequencies, and sources
- Rapidly changing technology—requiring an ability to keep up
- Documentation access and management—standards, SOPs, SWPs, tools, and templates
- Environmental and sustainability concerns
- Reputation management
- Weather and external conditions
- Business environment (BE) and economic factors leading to multiple unknowns.

Prequalification Challenge

Owners' internal and external prequalification processes often present major challenges and constraints to contractors. More importantly, in its current state, external prequalification does very little to improve HSE performance.

External prequalification provides an effective due diligence activity for owners to claim they have verified that invited contractors have met predetermined basic requirements. Among these requirements is the contractor's demonstration of the following:

- Ability to secure liability insurance coverage
- No use of child labor
- Compliance with the owner's corruption policies
- Possession of an HSE manual and safe work practices
- An HSE policy
- Ability to investigate incidents
- Historical HSE performance records

There is little (if any) verification of the quality of information provided by the prequalification service provider to satisfy prequalification requirements.

Technology Challenge

Most small- and medium-sized contractors do not possess adequate technology to monitor and manage the performance and ability of

DOI: 10.1201/9781003728467-5

their workforce to deliver safe, reliable, and efficient work performance. Contractors are challenged by the following in the acquisition of HSE management software to support performance management:

- Cost and budget constraints – most software providers offer an all-or-nothing model at extremely high cost (>300K USD acquisition and an equally high annual renewal cost).
- Irrelevant software modules
- Technology may be:
    - Complicated, and often not user-friendly
    - Perceived as a checkbox approach to monitoring HSE performance

Workforce Challenges

A contractor's workforce is generally transient. This is particularly so for small- to medium-sized contractors. Contractor workers are always seeking opportunities for greater stability, permanent employment with benefits, and opportunities for better growth and developmental potential. Furthermore, owners are always seeking highly skilled and competent workers to bolster their workforce for better performance.

Consequently, contractor organizations are vulnerable to turnover, and the cost of continuously training a new workforce is always challenging.

Common Challenges Faced by Contractors

General Challenges Experienced by Most Contractors

Contractors face a variety of other challenges depending on their industry and specific role. Addressing each of these challenges requires proactive planning, proper risk management strategies, and effective communication. If left unaddressed, these challenges often result in performance management concerns and issues. Common challenges are detailed below.

Challenge	Details
Budgets	• Contractors often struggle to stay within the project's budget due to unforeseen costs or inaccurate initial estimates. • Managing cash flow effectively to ensure that bills are paid on time while also maintaining profitability can be tricky, especially for smaller contractors with infrequent SOWs and employment.
Schedules and Deliverables	• Weather, unforeseen site conditions, and delayed material shipments can cause delays in completing projects on schedule. • Contractors frequently face pressure of meeting deadlines, which can affect work quality or lead to burnout with accompanying delays in schedules and deliverables.

Access to Skilled Labor	• There's often a shortage of skilled labor, which makes it difficult to find qualified workers for specialized tasks. • Contractors can face high employee turnover, which can lead to inconsistent quality and project delays. • Skilled labor prefers to work with owners and will often defect to owners' employment and staff.
Health and Safety Compliance	• Contractors are responsible for ensuring that they comply with HSE regulations, which can be complex and are constantly evolving. • Protecting workers from accidents and handling any safety violations is always a concern, especially on high-risk projects. • For small contractors in particular, a single event may lead to bankruptcy; injury frequencies can be insurmountable for meeting owners' threshold frequencies for employment. ○ Low hours worked lead to high TRIF / CRIF values.
Scope Creep and Changing Expectations	• Owners often change the scope of the project, whether it's adding new work or altering existing plans. ○ This can lead to additional costs and delays and is often uncompensated for relationship management. • When expectations aren't clearly defined from the start, contractors may find themselves working outside the original SOW. • Contractors often face changes in designs, materials, or timelines by owners that can cause additional costs or delays. • Owners may have unrealistic expectations about what can be achieved within a given timeframe or budget.
Quality Management	• Contractors must ensure that the quality of work meets both their standards and the client's expectations, which can be challenging with tight deadlines and limited resources. • Working with subcontractors who may not meet the same quality standards can affect the overall quality of work.
Reputation and Relationship Management	• Contractors often rely heavily on word-of-mouth and repeat business, so a negative review or a dissatisfied representative of the owner can harm their reputation. • Building and maintaining trust with owners, suppliers, and subcontractors is crucial for long-term success.

Uncontrollable Events	• Bad weather can cause significant delays, especially for outdoor construction projects. • Contractors may need to navigate challenges brought on by natural disasters like floods, hurricanes, or earthquakes. • Economic downturns or shifts in the housing market can reduce demand for construction services, while inflation or rising material costs can reduce profit margins. • Owners may struggle to secure financing, change their minds, and be subject to uncontrollable events (like a pandemic), leading to delays or cancellations of projects.
Technology Deficiency	• Keeping up with new software or technology for project management, estimating, or scheduling can be overwhelming for some contractors. • Contractors may struggle with ensuring the security of sensitive project data when using cloud-based tools or other digital platforms.
Prequalification	• Prequalification is seen as a task that brings little value to the contractor's HSE performance. • Contractors often farm out prequalification to third parties and pass the cost back to the owner. • For small- and medium-sized contractors, one incident can put the contractor out of business because of the low number of hours worked and accompanying high injury frequency.

Common Operating Challenges

Operating challenges that are common to most contractors during the execution of SOWs include the following:

Challenge	Details
Lack of HSE Training	• Contractors may not always receive adequate HSE training for the SOW and related activities. ○ This may include not only the general safety protocols but also the specifics related to the tasks they're performing or the equipment they're using. • Training regarding site-specific hazards and risks may also be general and of awareness level training • General weak awareness of organizational life-saving rules (LSRs), golden rules, and associated requirements.

Inadequate Personal Protective Equipment (PPE)	• PPE is expensive and requires frequent care, maintenance, and replacement. 　○ Contractors may not have access to or may not wear the appropriate PPE for the tasks they're performing, which can lead to HSE events. • PPE that can be used improperly ranges from helmets and safety goggles to fall protection harnesses. 　○ Improperly used PPE exposes the worker to hazards.
Working at Heights	• Falls from heights are a leading cause of fatalities and workplace injuries. 　○ Many contractors work at elevated heights, such as on scaffolding, ladders, or roofs, which exposes them to the risk of falls. • Ensuring proper safety protocols and equipment, like guardrails or fall arrest systems, is crucial.
Exposure to Hazardous Materials	• Contractors may encounter dangerous materials, such as toxic gases (e.g., carbon monoxide), liquids (e.g., diesel fuel), and respirable particulates (e.g., asbestos). • Failure to use proper precautions and PPE can lead to serious injuries and health risks.
Machinery and Equipment Hazards	• Contractors often work with heavy machinery, power tools, and other equipment. 　○ Improper use, maintenance, or lack of proper training on these tools can lead to accidents and/or injuries. • Machinery and equipment are costly and expensive to maintain and replace; consequently, they are often utilized well beyond normal operating lives to minimize cost. • With equipment life extension comes an increasing level of risk and vulnerability.
Fatigue	• Long working hours and repetitive strenuous tasks can lead to fatigue, which can compromise safety awareness and reaction times—making accidents more likely. • Contractors operate on the principle of *making hay while the sun shines* and will work consecutive long days over extended periods, exposing their workforce to fatigue, hazards, and risks. • With long, hot summer days and increasing obesity among workers, the risk of cardiac-related events is heightened. 　○ Vigilance in monitoring overexertion is essential.

Lack of Communication	• Effective communication is critical on a construction site. 　○ If owners and contractors don't communicate clearly with each other about hazards and risks, work processes, or changes in plans, conditions can result in accidents. • The contractor's workforce may be made up of a geographically diverse workforce with multiple languages—supervision in every worker's native language is not always available.
Weather Conditions	• Working in extreme heat, cold, rain, or high winds can create safety hazards. • Contractors must be trained to recognize the risks associated with different weather conditions and take necessary precautions.
Slip, Trip, and Fall Hazards	• Uneven surfaces, wet or slippery conditions, clutter, and improper footwear can all contribute to slip, trip, and fall incidents, which are some of the most common injuries on construction sites. • Slips trips and falls, a generally a leading cause of workplace injuries and fatalities.
Electrical Hazards	• Contractors often work around electrical wiring or power sources. • Failure to properly de-energize circuits or use insulated tools can result in electrocution or severe burns.
Inadequate Site Safety Management	• Sometimes safety protocols are not adequately enforced or maintained at the job site, leading to events. • Without a designated safety officer or clear safety rules, contractors may be at risk.
Lack of Emergency Procedures	• In case of an accident or emergency, having a clear plan for evacuation, first aid and communication is vital. • A lack of preparedness can worsen the outcome of an incident.
Environmental Hazards	• Exposure to extreme weather, poorly ventilated areas, or dangerous terrain can lead to accidents or health issues like heat stroke, hypothermia, or respiratory problems.
Manual Handling Risks	• Lifting, carrying, or moving heavy objects can lead to musculoskeletal injuries or strains, especially if proper techniques or equipment (like lifting aids or hoists) aren't used. • Pride often prevents workers from seeking help in manual handling and lifting activities.

Psychological Stress	• High-stress levels, pressure to meet deadlines, and dealing with difficult working conditions can contribute to accidents. • Mental health plays a significant role in maintaining overall safety on a job site.
Poorly Maintained or Unstable Structures	• In some cases, contractors work on buildings or structures that are old, unstable, or not up to code. • This can result in sudden collapses, debris falling, or other risks.
Leadership and Supervision	• Cost control can often result in a paucity of supervision and leadership during the execution of work, leading to costly events.
HSE Performance Data Analysis	• Except for large, mature contractors, HSE performance data is inadequately analyzed for common themes and proactive intervention.

Contractor Safety Management Challenges Faced by Owners

Contractor Evaluation and Selection

Challenges faced by owners in managing contractor safety include the following:

- Most owners do not have a well-established process for contractor evaluation and selection.
- Owners frequently face extensive paperwork and review requirements during contractor selection. This process overwhelms HSE teams, resulting in fewer contractors being invited to bid and, ultimately, possible suboptimal selection outcomes due to the review workload.
- Contractors often present extensive submissions to the SOW owner, including an HSE manual and information which is not easily verifiable—thus requiring time-consuming clarifications and interviews, which may overload the owner HSE teams.
- Desktop and field audits are infrequently performed despite their high value in performance improvements and contractor selection, particularly for repeat contractors.
- Contractor data analysis and information is difficult to analyze—there is no single repository, even within established prequalification providers.
- The process can be quite costly when many contractors are invited to bid on SOWs.

Business Environment Changes

Business environment changes affecting the owner and contractor relationship and work performance are detailed below.

- Fast-paced, low margin projects often lead to cost-cutting strategies that may impact HSE oversight and training budgets.

    - This happens in both the contractor and owner operating environments.
- Continuously evolving technological environments may impact the way work is performed.
- Artificial intelligence (AI) is fast becoming a source of truth for many organizations.
    - AI is still evolving. While it may be an excellent source of information, in the absence of field experience, major deficiencies may exist in AI-generated solutions.
- Decision-making speed has increased as organizations seek to remain competitive.
- The workplace is quickly being inherited by younger, educated but less experienced workers.
- Budgetary constraints often lead to cuts in HSE and training budgets where returns on investment are difficult to quantify and justify.

Mergers and Acquisitions

While not a new phenomenon, the business environment continues to be characterized by mergers and acquisitions. Such growth leads to an increased dependence on larger contractors and a dwindling dependence on small- and medium-sized contractors.

Communication Methods – Owner and Contractor

Modern virtual communication methods create greater dependence on informal and impersonal interactions between owners and contractors.

Such communication methods have both improved and contributed to failures in the communication process. Communication failures often include the following:

- Requirement and expectation details are often lost in verbal and non-documented communication processes.
- Body language, cognitive understanding, and verifications can be lost in a virtual communication process.
- The value of *showing* (the most effective process for communicating risks and hazards) is lost in the communication process.

Major Changes Affecting Owner Organizations

Within the past five years, many organizations (owners and contractors—more so owners) have undergone major changes that have impacted the way work is performed.

Major changes impacting the global workforce are detailed below:

- Pandemic-associated impacts
- Artificial intelligence (AI) impacts on knowledge generation
- Physical, evolving, environmental, and climatic changes
- Global trade realignment and tariffs

Year	Major Change
Global Change 2019	• Start of Covid-19 and ensuing lockdowns in 2020 • Geopolitical relationships and speeds of communication • Quality of communication
Industry Change 2020	COVID-19 work from home: start of a virtual and blended work environment. • Virtual training and communication • Only essential services and workers on worksites • Little field training, if any • Reduction in new and sustaining projects • Mobile workforce
Organizational Change 2021	Restructuring: leaner and semi-decentralization of HSE Org. • Centralized Operational Risk Management (ORM) organization • Layoffs / retirements / relocations – loss of HSE experience and expertise. • Little HSE training in the field (even less in the contracting workforce) • No major projects (migration of skilled labor)
Organizational Change 2022 and Ongoing	Large scale return to work, uptick in project work – turnaround and sustaining projects • Issues with the workforce being insufficiently trained and competency assured • Lean workforce (*Is it agile?*) • Somewhat decentralized EH&S support (*Is it balanced?*) • Overwhelmed workforce • Oil prices / tariffs drive work volumes and performance • AI is seen as a means of generating documented knowledge and reducing the cost associated with work practices
Workforce Generational Changes	The workforce generational challenge is characterized as follows: • A decreasing number of field-experienced Gen X workers • An increasing number of younger, inexperienced Gen Y / Z workers in the workplace who learn differently from Gen X and are generally deficient in field experience • An aging workforce with workers remaining a longer time in the workforce. Those who remain may find difficulties in adapting to technological changes and may be at higher risk of health and safety events such as musculoskeletal injuries, slips, and falls.
World Trade – 2025	In early 2025, tremendous uncertainty was introduced into domestic and global markets because of world trade realignments and the introduction of tariffs and trade restrictions. These measures have impacted both the contractor and employee workforces.

Contractor—Size and Related Attributes

Size and Related HSE Attributes

Over several years of observation, analysis, and documentation, we have summarized attributes of contractor organizations relative to their size. While not an exact science, presentations and feedback across the world—including Canada, the USA, the Middle East, Asia, and the Caribbean—allowed us to compile data relative to HSE attributes and size relationships of contractor organizations as shown in Figure 5.1.

- Highlighted items reflect the contractor's capabilities relating to HSE and the potential for the contractor to deliver safe, reliable, and efficient work performance.
- Within each group of contractors, HSE attributes varied based on the leaders' understanding, knowledge, and care for the workforce.
- The information provided in this summary is directional and helps leaders better understand the challenges and opportunities within each contractor group to enhance HSE performance and the delivery of safe, reliable, and efficient work.
- These attributes also help owners establish proactive programs to improve HSE performance in the workplace.

< 10 Employees	11-30	31-100	101-1000	1001 -3000	>3000
• Relatively safe • Fully supervised by owners • Follows owner processes and procedures • Spend time building brand - very careful to avoid incidents • **Very little knowledge / focus on HSE and HSE MS** • Can go out of business with one incident	• Vulnerable because of growth focus • **New to owners so most work supervised by owners** • **Growing focus on HSE - No HSE MS** • Follows owner processes and procedures • Can go out of business with one incident	• Vulnerable because of growth • **Inexperience / Untrained workforce** • Growing SMEs so generally allowed to perform work unsupervised by owners • Trying to manage growth and HSE performance	• Improving HSE performance • Performs work under contractor supervision • Follows owner processes and procedures • Try hard to avoid incidents • **Excel driven HSE MS / processes and tools**	• Relatively safe • Supervised by Contractor organization • Little owner supervision • Newly acquired HSE MS and tools • **Growth through mergers and acquisitions ... consolidating HSE MS**	• Relatively safe • Mature HSE MS and culture • Improving processes • Some best-in-class industry performance, processes and tools • **Functions with owner oversight only ... inconsistent auditing**

Figure 5.1: HSE attributes and size relationships of contractor organizations
Source: OEMS Erudite Corp. (2025).

How Owners Can Help Improve Contractor HSE Performance

Addressing the Challenges Faced by Contractors

Delivering business performance excellence is a joint expectation from both owners and contractors. Beyond the performance management requirements described in Chapter 4, a major limiting factor in improving contractor HSE performance is their access to performance-enabling tools and processes.

Owners can better enable contractors in the delivery of excellence in SOWs by helping in the following ways:

- Providing access to quality, usable documents and documented requirements, allowing contractors and users to easily interpret, apply, and deliver the owner's requirements.
- Supporting contractors with fit-for-purpose HSE management systems.
- Ensuring access to performance enablement tools and processes for effective stewardship of HSE performance. See Figure 5.2 for integrated HSE enablement tools and processes.
  - Required performance enablement tools and processes (software modules) are aligned with elements of an integrated management system (outer circles) and the PDCA Model.
- Providing specific training to contractors based on identified weaknesses and opportunities to prevent events.
- Implementation of proactive interventions where appropriate from data analyses and industry learning.

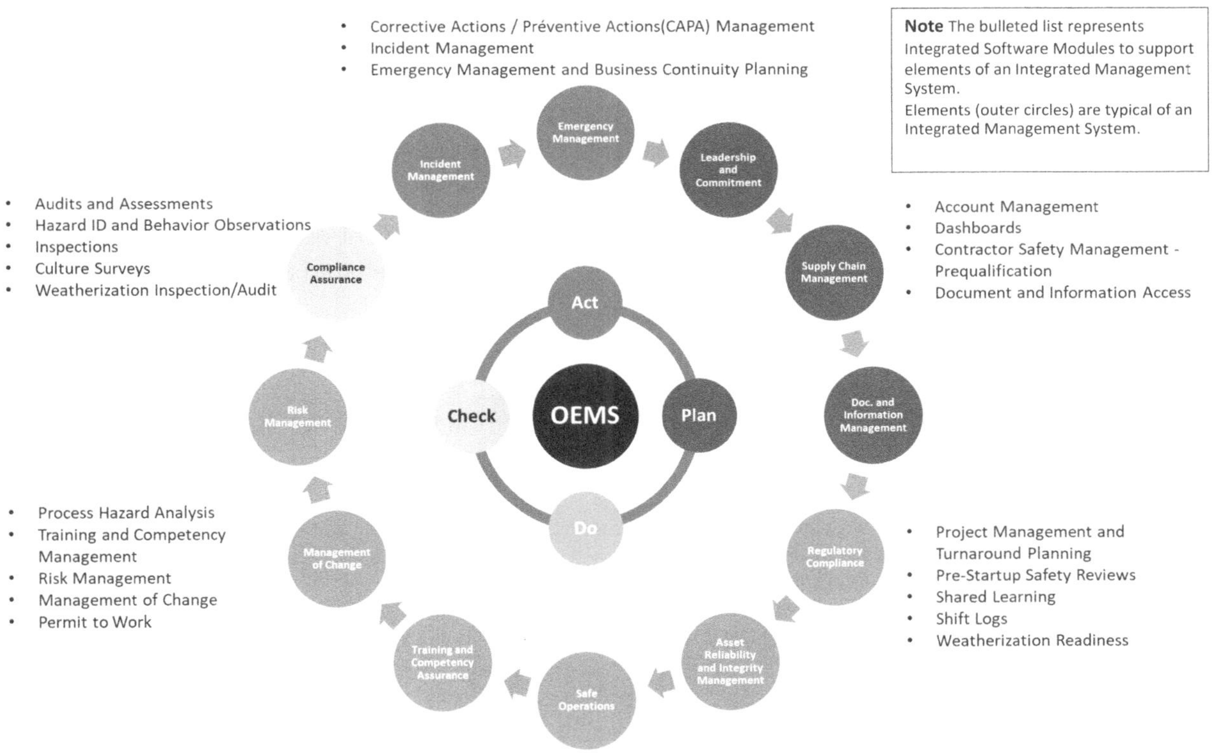

Figure 5.2: Integrated technological tools and processes for HSE enablement.
Source: OEMS Erudite Corp. (2025).

Enablement Tools for Improving Contractor HSE Performance Based on Contractor Size

Startup Small Contractors <10 Employees

Small contractors are generally in the early stages of business startup, usually more focused on delivering technical expertise in a particular field. They are:

- Generally unaware of the value of an HSE management system and, in many instances, unaware of documented safe work practices and processes to manage HSE performance.
- Equipped with an immature workforce and has not yet gelled as a smoothly functioning team.
- Deficient in HSE resources and support requirements.
- Vulnerable when performing field work, requiring the support and direct supervision of owners to perform work safely.
- Willing to learn and grow their business and will meet any requirement imposed upon them by the owner.
- Unable to purchase HSE management technology and will often depend on paper systems to meet the owner's reporting requirements.

These conditions present an excellent opportunity to shape the safety culture of these small contractors. Owners can enable HSE excellence in small contractors by supporting their access to, and use of, the various HSE enablement tools detailed in Figure 5.3. Contractors also benefit immensely from HSE training mandated by or provided by the owner for safe, reliable, and efficient delivery of work.

Such contractors aim to meet every requirement of the owner for the safe delivery of work as they seek to become a preferred contractor, thereby growing their business.

< 10 Employees	Owner can enable the contractor's workforce with the following tools and processes:
• Relatively safe • Fully supervised by owners • Follows owner processes and procedures • Spend time building brand - very careful to avoid incidents • **Very little knowledge / focus on HSE and HSE MS** • Can go out of business with one incident	• Incident Management – Recording through Investigation and Root Cause Analysis (RCA) • Document Access – As and when Required (Policies / Standards / Procedure/ SWPs) • Inspections - Daily / EH&S / Leadership • Corrective Actions / Preventive Actions Management (CAPA) • Shared Learning (EH&S Alerts and Best Practices / Toolbox Talks / Safety Moments & SWPs) • EH&S Training and Competency Management – Training Matrix Owner can help the Contractor in developing its HSE MS and deliver periodic HSE Training to its workforce.

Figure 5.3: Suggested enablement tools for startup small contractors.
Source: OEMS Erudite Corp. (2025).

Growing Small Contractors (11–30) Employees

Growing small contractors are generally past startup and in the early stages of business growth. They are:

- More established in the delivery of a technical expertise—and doing it well.
- Growing in HSE awareness with a better understanding of PPE requirements, safe work practices, procedures, and documented approaches to doing work.
- Able to maintain a dedicated HSE resource, procured as required to meet the owner's requirements.
- Not yet able to attract skilled and competent workers who tend to be wary of layoffs when projects are scarce—and not yet willing to become full-time organization employees.
- Equally as vulnerable as small contractors and therefore require similar support and treatment by the owner.

Owners can enable HSE excellence in small, growing contractors by supporting their access to and use of the various HSE enablement tools detailed in Figure 5.4. They may also benefit from document sharing by the owner and from site-specific technical training as appropriate.

11-30	Owner can enable the contractor's workforce with the following tools and processes:
• Vulnerable because of growth focus • New to owners so most work supervised by owners • Growing focus on HSE - No HSE MS • Follows owner processes and procedures • Can go out of business with one incident	• Incident Management – Recording through Investigation and Root Cause Analysis (RCA) • Document Access – As and when Required (Policies / Standards / Procedure/ SWPs) • Inspections - Daily / EH&S / Leadership • Corrective Actions / Preventive Actions Management (CAPA) • Shared Learning (EH&S Alerts and Best Practices / Toolbox Talks / Safety Moments & SWPs) • EH&S Training and Competency Management – Training Matrix Owner can help the Contractor in developing its HSE MS and deliver periodic HSE Training to its workforce. Owner can also provide appropriate SWPs.

Figure 5.4: Suggested enablement tools for growing small contractors.
Source: OEMS Erudite Corp. (2025).

Medium-Sized Contractors (31-100) Employees

Medium-sized contractors are in a rapid growth phase and generally maintain a full pipeline of work opportunities. They are:

- Prepared to accept any type of work and are vulnerable for this primary reason. The high growth rate generally results in a largely untrained and inexperienced workforce.
- Large enough to supervise themselves and with growing trust by the owner. They are frequently considered for new types of work based on prior performance evaluation, fully accountable for delivering excellent work.
- Able to more easily attract and retain a skilled and competent workforce; the workforce is maturing in tenure and skills.

- Generally, still unable to purchase highly valuable performance improvement integrated management systems tools and processes due to cost constraints—however, they are more inclined to do so.
- Now include a dedicated HSE team to provide HSE oversight of the workforce during the execution of work.

Owners can enable HSE excellence in growing medium-sized contractors by supporting their access to and use of the various HSE enablement tools detailed in Figure 5.5.

31-100	Owner can enable the contractor's workforce with the following tools and processes:
• Vulnerable because of growth • **Inexperience / Untrained workforce** • Growing SMEs so generally allowed to perform work unsupervised by owners • Trying to manage growth and HSE performance	• Incident Management – Recording through Investigation and Root Cause Analysis (RCA) • Document Access – As and when Required (Policies / Standards / Procedure/ SWPs) • Inspections - Daily / EH&S / Leadership • Weatherization – Extreme weather conditions • Process Hazard Analysis (PHA) - HAZOP / FMEA / What if –Checklist / FTA • Corrective Actions/Preventive Actions Management (CAPA) • Permit to Work • Pre-Startup Safety Reviews (PSSR) • Risk Management (Operations / Projects and Enterprise) – Consistent with Suncor Risk Matrix • Shared Learning (EH&S Alerts and Best Practices / Toolbox Talks / Safety Moments & SWPs) • EH&S Training and Competency Management – Training Matrix • Audits (ISO 45001 / PSM / OEMS / API 1173 and other Management Systems) • Management of Change (MOC) • Dashboards – Trending and Analysis.
	Owner should provide targeted HSE Training to Contractors.

Figure 5.5: Suggested enablement tools for growing medium-sized contractors.
Source: OEMS Erudite Corp. (2025).

Mature Medium-Sized Contractors (101–1000) Employees

Mature medium-sized contractors generally maintain a trained and competent workforce and perform work safely, reliably, and efficiently. They:

- Maintain effective management system labor intensive tools and processes, generally Excel-driven.
- Form an integral part of the owner's business and are required to have a well-developed, integrated management system with supporting tools and processes.
- Are in the early stages of acquiring software tools and processes.
- Can be enabled by owners for HSE excellence through supported access to and use of the various HSE enablement tools detailed in Figure 5.6.

Such contractors can benefit immensely from contractor safety forums for learning opportunities from peers and from targeted safety workshops.

Growing Large Contractors (1001–3000) Employees

Growing large contractors generally maintain a trained and competent workforce and perform work safely, reliably, and efficiently. They are:

- Able to maintain effective management system labor intensive tools and processes, generally Excel-driven.

101-1000	Owner can enable the contractor's workforce with the following tools and processes:
• Improving HSE performance • Performs work under own supervision • Follows owner processes and procedures • Try hard to avoid incidents • **Spreadsheet (Excel) driven HSE MS / processes and tools**	• Incident Management – Recording through Investigation and Root Cause Analysis (RCA) • Document Access – As and when Required (Policies / Standards / Procedure/ SWPs) • Inspections - Daily / EH&S / Leadership • Weatherization – Extreme weather conditions • Process Hazard Analysis (PHA) - HAZOP / FMEA / What if –Checklist / FTA • Corrective Actions/Preventive Actions Management (CAPA) • Permit to Work • Pre-Startup Safety Reviews (PSSR) • Risk Management (Operations / Projects and Enterprise) – Consistent with Suncor Risk Matrix • Shared Learning (EH&S Alerts and Best Practices / Toolbox Talks / Safety Moments & SWPs) • EH&S Training and Competency Management – Training Matrix • Audits (ISO 45001 / PSM / OEMS / API 1173 and other Management Systems) • Management of Change (MOC) • Dashboards – Trending and Analysis. Owners should consider providing: Annual Contractor HSE Conferences (Forums) and Targeted HSE Workshops.

Figure 5.6: Suggested enablement tools for mature medium-sized contractors.
Source: OEMS Erudite Corp. (2025).

- Form an integral part of the owner's business and are required to have a well-developed employer aligned integrated management system and supporting tools and processes.
- In a high growth rate phase from mergers and acquisitions:
  - Growth through mergers and acquisitions requires integration periods of multiple cultures within the organization, increasing the risk of major events.

Owners can enable HSE excellence in growing large contractors by supporting their access to and use of the various HSE enablement tools detailed in Figure 5.7.

1001-3000	Owner can enable the contractor's workforce with the following tools and processes:
• Relatively safe • Supervised by Contractor organization • Little owner supervision • Newly acquired HSE MS and tools • **Growth through mergers and acquisitions ... consolidating HSE MS**	• Incident Management – Recording through Investigation and Root Cause Analysis (RCA) • Document Access – As and when Required (Policies / Standards / Procedure/ SWPs) • Inspections - Daily / EH&S / Leadership • Weatherization – Extreme weather conditions • Process Hazard Analysis (PHA) - HAZOP / FMEA / What if –Checklist / FTA • Corrective Actions/Preventive Actions Management (CAPA) • Permit to Work • Pre-Startup Safety Reviews (PSSR) • Risk Management (Operations / Projects and Enterprise) – Consistent with Suncor Risk Matrix • Shared Learning (EH&S Alerts and Best Practices / Toolbox Talks / Safety Moments & SWPs) • EH&S Training and Competency Management – Training Matrix • Audits (ISO 45001 / PSM / OEMS / API 1173 and other Management Systems) • Management of Change (MOC) • Dashboards – Trending and Analysis. Owners should consider providing: Annual Contractor HSE Conferences (Forums), Targeted HSE Workshops, Oversight and Aligned Integrated Management System.

Figure 5.7: Suggested enablement tools for growing large contractors.
Source: OEMS Erudite Corp. (2025).

Such contractors can benefit immensely from contractor safety forums for learning opportunities from peers and from targeted HSE workshops and owner-aligned integrated management system expectations and requirements.

Mature Large Contractors (>3000) Employees

Mature large contractors generally maintain a trained and competent workforce and perform work safely, reliably, and efficiently. They are:

- Able to maintain effective integrated management systems that may be more mature than the owner's.
- Subject to detailed owner audits and assessments per Figure 5.8 and may be subject to owner oversight as required.
- Able to share best practices and deliver leading-edge technologies and work practices to the owner.
- Generally willing to be lead participants and sponsors of HSE forums and learning sessions.

>3000	Owner can enable the contractor's workforce with the following tools and processes:
• Relatively safe • Mature HSE MS and culture • Improving processes • Some best-in-class industry performance, processes and tools • **Functions with owner oversight only ... inconsistent auditing**	• Incident Management – Recording through Investigation and Root Cause Analysis (RCA) • Document Access – As and when Required (Policies / Standards / Procedure/ SWPs) • Inspections - Daily / EH&S / Leadership • Weatherization – Extreme weather conditions • Process Hazard Analysis (PHA) - HAZOP / FMEA / What if –Checklist / FTA • Corrective Actions/Preventive Actions Management (CAPA) • Permit to Work • Pre-Startup Safety Reviews (PSSR) • Risk Management (Operations / Projects and Enterprise) – Consistent with Suncor Risk Matrix • Shared Learning (EH&S Alerts and Best Practices / Toolbox Talks / Safety Moments & SWPs) • EH&S Training and Competency Management – Training Matrix • Audits (ISO 45001 / PSM / OEMS / API 1173 and other Management Systems) • Management of Change (MOC) • Dashboards – Trending and Analysis.
	Owners should consider providing: Annual Contractor HSE Conferences (Forums), best in class SWPs, Oversight and Audits.

Figure 5.8: Suggested enablement tools for rapidly mature large contractors.
Source: OEMS Erudite Corp. (2025).

6 CONCLUSION

Overview

There is no single silver bullet for improving contractor safety management. Sustainable contractor HSE performance improvements require accurate and consistent interpretation and application of well-designed processes and protocols.

Among these are the following:

- Joint leadership commitment and accountability—it's no longer an "us and them" concern. Both parties benefit immensely from joint commitment to improving contractor safety performance.
- Effective contractor prequalification—our historical approach to external prequalification is ineffective and adds to the workloads of contractors who now see prequalification as a checkbox exercise to getting work with an owner.
- Onboarding, training, and competency assurance—effective onboarding (premobilization/mobilization) and competency assurance of the contractor workforce are very impactful in the delivery of performance excellence.
- Effective communication—careful choice of methods, media, and frequency is important when communicating knowledge to the workforce.
- Effective supervision—work supervision cannot be underestimated in the delivery of safe, reliable, and efficient work. Contractor and owner supervision is required for HSE performance improvements. Supervision may be costly, but it is extremely effective.
- Audits, compliance assurance, performance management—the value of audits, assessments, compliance assurance and verifications, and contractor performance management is paramount to success in CSM.
- Digitalization and data analytics—contractors are cost constrained when attempting to acquire commercial tools and processes for digitalization and data analytics and must be better enabled with support from owner companies.
- Fair treatment and collaboration—treating contractors fairly and collaborating on finding the safest ways to perform work provides opportunities for enhancing contractor safety performance.
- Risk-based management approach—managing contractor work based on the risks presented can improve the delivery of safe, reliable, and efficient performance.

Focusing on the Big Picture

Within my *organization*, there may be no contractor fatalities or contractor events. Within my *industry*, however, there generally are multiple SIFs and contractor events. Leaders should ask themselves, "Are we doing something really well, or are we just lucky to evade SIFs?"

Leaders should be able to attest that sufficient evaluation and analysis of events within the organization have been conducted to identify

DOI: 10.1201/9781003728467-6

commonalities that may be indicative of underlying, untreated SIF and event precursors.

Effective owners acknowledge the following:

- We can do a better job of analyzing HSE performance data beyond communicating information to leaders.
- The underlying information useful for event prevention generally does not surface—unless we look for it.
- An absence of recent SIFs can create a false sense of security that everything is well and we are on top of things.
- Contractor SIFs are indicators of HSE performance and are undervalued when we treat the owner-contractor relationship as *us* and *them*.

Understanding the bigger picture within my organization (across business units and business areas), within my industry, and across industries helps shape the response strategy for contractor HSE performance improvements. HSE professionals and leaders should continuously scan the business environment for HSE concerns, SIFs, and events. Failure to do so increases organizational vulnerability and fails to identify proactive measures for improving HSE and contractor safety performance.

Insufficient Data Analysis

Leaders should ensure that data analysis and evaluations are ongoing and track key metrics. Total recordable injury frequencies (TRIF), employee recordable injury frequencies (ERIF), and contractor recordable injury frequencies (CRIF) are calculated numbers used to provide spot information and trends over time to leaders. A falling trend is seen as a good thing.

However, stratified analysis beyond this allows us to better understand common themes and root causes. Categories include:

- Age of workers involved in the event (e.g., young and inexperienced)
- Injury source (e.g., life-saving rules violations)
- Business location (i.e., Is one business area more impacted than others and, if so, why?)
- Work activities (i.e., Are we experiencing SIFs in low and moderate risk work activities and, if so, why?)
- Season (e.g., winter, where the likelihood of seasonal depression is increased; summer, where extended daylight and extreme heat may precipitate overwork and heat-related injuries)

Trend analyses are limited when events are infrequent and occur over multiple years in different business locations.

When leaders focus on the big picture over extended periods, a clearer picture evolves that enables the application of proactive intervention for business performance excellence.

Shift Focus to Excellence in Performance

Instead of demanding contractor work to be performed ***cheaper, better, and faster***, owners should consider shifting their approach to contractor management that emphasizes ***safe, reliable, and efficient*** (***excellence***) delivery of work. The demand for cheaper, better, and faster should not be part of the vocabulary of leaders in the procurement and contractor management disciplines. Excellence in performance is relatable to all leaders and is much more effective in improving contractor safety performance.

Such a shift requires that the entire leadership of the organization becomes competent in HSE leadership and able to relate to the phrase: ***Good Safety Means Great Business***. This means that CFOs and budget managers/non-field leaders must understand that investments in training and competency management, HSE, and contractor safety management have an unquantifiable yet hugely positive return on investment—and this should not be underestimated. With this knowledge, budget cuts that adversely impact these variables should be less frequent, shallower, and perhaps eliminated from the business planning process.

Leadership—Genuine Care and Empathy

Treating contractors with genuine care and empathy is extremely effective in the delivery of work excellence. The litmus test for the execution of all work performed by both employees and contractors is this the response to this question: "Would I permit my loved one (child, brother, sister, relative) to do this work with the current controls?"

If the answer is no, then the current controls are more than likely insufficient.

Unless leaders really stand behind slogans like:

- If we can't do it safely, we will not do it at all
- No work is so urgent or important that we can't take the time to do it safely
- Zero harm and target zero

There will always be events and—most often—contractors are the ones to suffer.

When we are capable of doing so and can adequately respond to the pressures of higher levels leadership, we have genuinely taken a giant step toward performance excellence on the journey to improvements in contractor safety management.

Contractor Enablement—Tools and Processes

External contractor prequalification processes continue to be a checkbox exercise imposed on contractors to enable them to bid on work offered by an owner. What is more important than external prequalification is an owner's ability to easily review the HSE performance of a contractor as required. At present, this ability is quite limited because contractors do not have tools and processes that are easily accessible and reasonably priced.

The authors believe that enabling and supporting contractor's access to and application of HSE management tools and processes is more beneficial than external prequalification in the improvement of HSE performance. Alternatively, current external prequalification service providers must evolve to better address the challenges faced by contractors in their quest for delivering safe, reliable, and efficient work.

Closing Statement

Contractor safety management is perhaps the best opportunity for owners to improve HSE performance. There are many unexplored and underutilized opportunities for improving contractor HSE performance. Harnessing these opportunities requires bold and effective leaders who are prepared to work with contractors and address the many challenges they face, as described in Chapter 5.

This is a collaborative effort between both owner and contractors with a genuine desire to create win-win outcomes and HSE performance improvements. Sustained improvements require the joint contractor and owner commitments to the following:

- Giving the contractor safety management process time to work
- Enabling contractors with workable tools and processes
- Removing punitive consequences for reporting events and gaps in compliance
- Applying a prequalification process that adds value to the contractor safety management process
- Proactive learning from events and translating knowledge to the frontline
- Continuously learning and improving the process

It's important to remember that contractor safety management is a dynamic process which is constantly evolving with industry. Consider again the advent of seatbelts for vehicular safety, whose early years were marred by inconsistent use, poor design, and resistance to change. However, consistent effort and improvements created a safety culture where a seatbelt is now a reflex—rarely do we have to remind each other to *buckle up*. Continuous improvement in CSM is therefore not simply an administrative responsibility, but a moral necessity. We can keep improving and, in the process, save lives.

We would like to extend our sincere gratitude to you, the reader, for the time and attention required to engage with the material presented. It reflects your commitment to making meaningful progress in contractor safety, and for that we thank you.

APPENDIX 1 CONTRACTOR SAFETY MANAGEMENT STANDARD/ PROTOCOL

Owner Logo	Integrated Management System	
Document Title: **Contractor and Supplier Management Standard**		
Document ID:	**Version No.**:	**Total # of Pages**:
Legacy Doc: NA		

Comments:

- Element supporting documents and tools are available on the IMS SharePoint site under Element E8 Contractor and Supplier Management.
- Implement requirements and expectations detailed in this Standard using the full listing of the following:
    - Supporting documents and programs.
    - Execution support tools.
    - External references.

			Endorsement		Approval
Issued for Use		**Signature**			
		Name			
Status	**Date (YYYY-MM-DD)**	**Role**	**Element Subject Matter Expert**	**Administrator**	**Owner**

DOCUMENT USE DISCLAIMER:

To ensure that an approved revision of this document is being used:

- Confirm the Revision field contains a number greater than zero, no letters, and the Status indicates "Issued for Use."
- Users are responsible for ensuring that they are using the most current version as issued in the Electronic Document Management System.
- If the Revision field contains "99," or the Status field contains "Issued for Information Only" (or both), then the document is considered as reference only and carries no authority.

Note:

CONFIDENTIALITY	All rights are reserved. No part of this document may be reproduced or transmitted in any form or by any means without the written permission of the owner

	Version	Date (YYYY-MM)	Description/Text Affected	Remarks
REVISION LOG	1		Original publication	Issued for Use

Table of Contents

About This Document

Scope and Purpose

The scope of this Standard is contractor and supplier management and how the owner selects and manages contractors and suppliers for safe, reliable, and efficient performance.

The purpose of this standard is to describe the following:

- The lifecycle requirements of contractor and supplier management within the organization.
- Requirements and expectations regarding owner management of contractor and supplier selection and performance alignment with its management system.

Objectives

The objectives of this standard are as follows:

- Contractors and suppliers are selected using a methodology that considers the following:
    - Health, safety, environmental, and operational performance
    - Technical and operational capabilities
    - Local and in-country ownership
    - Partnerships and community involvement
    - Commercial competitiveness
- Contractor selection considers SOW risks and the competency of all entities engaged to safely perform the work.
- Management processes are in place to ensure that contractors are executing tasks consistent with the owner's expectations.
- CSM processes exist to engage with leaders at all levels of respective organizations.
- An approved list of contractors and suppliers is available to the business.

> **NOTE:** In this document, "contractors" refers to contractors *and* subcontractors performing hands-on work at owner worksites. "Suppliers" refers to organizations supplying equipment, machinery, and materials to the owner.

Expectations

Expectations associated with the owner's contractor and supplier management are as follows:

	Expectations
Prequalification and Selection	• The SOW owner prequalifies and selects contractors, maintaining an approved contractor list in alignment with owner expectations, prior to being awarded work.
Competency	• The owner assesses contractors' qualifications, experience, and competencies prior to awarding and commencing work.

Risk Management	• Controls are in place to manage contractor and supplier activity, promoting safe, reliable, and efficient performance.
Performance Management	• Processes are in place to promote materials and services that meet the owner's specifications. • Processes are in place to verify contractor adherence to owner expectations of safe, reliable, and efficient operations.

Application

This standard applies to selection and management of all contractors performing work for the owner and for suppliers providing materials for the owner's operations. This includes those under the owner's direct supervision and control and those assigned prime contractor ownership.

> **NOTE:** Where contractors are assigned prime contractor ownership, selection follows the requirements detailed in this standard.

Users

Users of this standard include all workers (employees and contractors) across the owner operations involved in selection and management of contractors and suppliers.

Core Supporting Documents and Processes for Meeting Requirements

Owner-Wide Policies

Owner-wide policies applicable to this standard include:

- Owner's Drug and Alcohol Policy
- Owner's Respectful Workplace Policy
- Code of Business Conduct and Ethics

Owner Practices

Practices required to meet requirements of this standard are as follows:

- Financial and legal requirements for contractors' practice
- Procurement practices

Operations-Wide Tools and Processes

Tools and processes required to meet requirements of this standard are as follows:

- Approved contractor and supplier lists
- Owner's contractor prequalification process
- Contractor HSE audit
- Contractor HSE inspections
- Owner's safe work practices
- Contractor safety improvement plan
- Request for Information/Proposal (RFI/RFP) process

Terms, Abbreviations, and Definitions

Terms and Definitions

Terms and definitions relevant to this standard are defined below. Additional definitions may be found in the IMS Dictionary.

Term	Definition
Category	A common grouping of materials or services.
Competitive Sourcing	A formal process whereby a request is made for multiple contractors or suppliers to submit offers for a particular material or service.
Contractor	Any contractor/subcontractor that will perform work on an owner worksite or on behalf of owner. **NOTE:** When suppliers are required to manufacture, deliver and/or install equipment and/or machinery on owner worksites, they are considered contractors.
Critical Contractor	A contractor deemed essential to operations due to the nature of their work, whose involvement presents a heightened risk of safety incidents. This designation requires additional communication and engagement to mitigate potential safety risks and ensure compliance with operational standards and expectations.
Critical Supplier	A supplier providing materials, equipment, and/or machinery crucial to operations—those directly affecting safety, reliability, availability, schedule, and cost.
Prequalified Contractors	Contractors that have attained approved prequalification status in the owner prequalification database.
Prime Contractor	The contractor who, by virtue of contractual terms, is responsible and accountable for all worksite activities. The owner is generally the prime contractor unless otherwise specifically stated.
Sole Source	Only one supplier can provide the commodity and/or perform the services required due to uniqueness.
Supplier	An organization or entity that provides equipment, machinery, and/or materials to owner.
Supply Chain Management (SCM)	The overarching processes and practices used to identify demand and procure goods and services for the organization. SCM is sometimes used interchangeably with Procurement.
Subcontractor	An organization or entity performing work on owner worksites on behalf of an owner contractor under a specific agreement with the owner contractor.

Abbreviations

Abbreviations used in this document are as follows:

Abbreviations	Definition
CSM	Contractor and Supplier Management
H&S	Health and Safety
HSE	Health, Safety and Environment
KPI	Key Performance Indicator
OE	Operational Excellence
PM	Project Manager
RFP	Request for Proposal
SCM	Supply Chain Management
SIP	Safety Improvement Plan
SOR	System of Records
SOW	Scope of Work
WCB	Worker Compensation Board

Roles and Responsibilities

Proficiency Levels Defined

Roles identified within this standard are required to maintain the appropriate competency and proficiency level to meet assigned responsibilities. Where applicable:

- Competency is developed through a combination of training and experience.
- Proficiency is the degree to which a worker is deemed competent, having received increasingly advanced training and more complex experience.

IMS Standard Training

IMS training and experience associated with this standard includes the following:

Training Level	Objective/Focus
Awareness	• Contractor and supplier management awareness training (basic level). • Awareness of basic contractor oversight processes, including onboarding and orientation processes. • Ability to participate in non-oversight roles such as procurement and data management.
Knowledge	• Understanding of the owner's procurement requirements and processes for contractors and suppliers. • Understanding of IMS requirements for contractors at an owner worksite.

	• Working knowledge of prequalification and evaluation tools for selection of approved contractors. • Ability to lead contractor pre-mobilization and onboarding at an owner worksite. • Ability to oversee contractors and/or suppliers using standard tools, including audits, inspections, and safety improvement plans.
Subject Matter Expert	• Ability to design and oversee workflows related to contractor management, including evaluation and performance management.
Mastery	• Mastery level application of contractor and supplier management, including all aspects of this standard.

> **NOTE:** The requirements of each proficiency level are in addition to the requirements of all proficiency levels that precede it.

Roles and Responsibilities

Roles and responsibilities associated with this standard are as follows:

Management System Roles and Responsibilities

Roles	Training Level	Responsibility
Standard Owner	Knowledge	Responsible for adequacy of contractor and supplier management expectations and requirements, and overall standard performance.
Standard Subject Matter Expert (SME)	Mastery	Responsible for content quality, standard expectations, functionality, supporting program and standard documents and system of record (SOR) effectiveness. External SMEs may be engaged to create and maintain IMS program and supporting documents when deemed necessary by owner management.
Management System Stewardship Department	Subject Matter Expert	Responsible for all management system oversight and administration. Responsible for assessing contractor and supplier management effectiveness, including awareness, competency, efficiency, and improvement across the organization. May be called upon to support contractor audits and support the development of contractor and supplier management awareness training as required to support the implementation of this standard.

Contractor and Supplier Management Standard-Specific Roles

Roles	Training Level	Responsibility
Supply Chain Management	Mastery	Supply chain management is responsible for: • Developing, implementing, and maintaining contractor and supplier procurement processes in alignment with the owner's management system. • Ensuring contractor and supplier evaluation processes are aligned and verified with the owner's management system requirements. • Assessing contractor and supplier management processes for adequacy and conformance. • Ensuring that management engagement activities occur between the owner and its contractors and suppliers.
Category Management	Subject Matter Expert	Category management is responsible for: • Performing work associated with the strategic approach to procurement. • Creating category sourcing strategies, strategy execution (RFP's & contract negotiations). • Making contract award recommendations. • Establishing pricing agreements. • Ongoing contractor and supplier management.
Procurement	Subject Matter Expert	Procurement team is responsible for: • Performing work associated with transactional procurement, including purchase requisitions and purchase orders. • Performing tactical sourcing events.
Contract Owner—Category Manager	Subject Matter Expert	The contract owner accountable for the contract is responsible for the following: • Understanding the requirements of this standard and ensuring these are considered in the scope of the contract. • Supporting the SOW owner—project manager in resolving administrative concerns with contractors or suppliers as they arise.

		• Being the administrative point of contact between the owner's worksite leadership and the contractor/supplier, facilitating concern resolutions when required. • Participating in the contract kick-off and supporting pre-mobilization activities as requested. • Verifying that selected contractors are prequalified; and, where not so, appropriate approvals and documentation accompany contractor selection.
SOW Owner—Project Manager	Knowledge	The SOW owner (project manager) is responsible for: • Following the owner's procurement requirements, ensuring all contractors and suppliers are prequalified and approved prior to use. • Assuring necessary oversight controls (including supervision requirements) are identified and in place prior to work commencing. • Assuring completion of all re-work, ongoing work, and post-work activities for contractors and suppliers, including critical contractors and critical suppliers. • Conducting periodic performance evaluations of contractors and suppliers where required. • Initiating the development of a safety improvement plan where required. • Holding contractors accountable for fulfilling the terms of agreements—including the requirement to follow owner's expectations while working on owner worksites.
Legal Support	Awareness	The owner's legal support is responsible for: • Reviewing and approving contract terms, deviations, and other legal counsel support when required.

General Company Roles

Roles	Training Level	Responsibility
Executives	Knowledge	Executives are responsible for the following: • Resourcing the development of effective contractor and supplier management processes. • Maintaining engagements with long-term strategic contractors and suppliers, including high-value contracts.
Managers	Knowledge	Owners' managers are responsible for: • Understanding the basic requirements for contractor and supplier management and ensuring compliance with those requirements. • Promoting early engagement with procurement as soon as a need for goods and/or services arises. • Ensuring SOW owners are competent in the scope of services requested. • Ensuring only approved contractors and suppliers are used for operations. • Developing and maintaining an active list of critical contractors and critical suppliers, and ensuring the enhanced controls are implemented for each category. • Maintaining active engagements, including field leadership visits and visibility during the execution of work (ideally with peer contractor leaders). • Monitoring contractor performance and supporting SOW owner project managers with holding contractors accountable for adhering to the owner's expectations. • Ensuring that oversight is in place and functioning, including the following: ○ Site-specific onboarding and orientation. ○ Minimum supervisory levels are maintained and aligned to the risk of tasks being conducted. ○ Quarterly meetings and engagements with applicable contractors. ○ Field visits during work execution. ○ License verification, and use of required procedures, practices, and tools.

HSE	Knowledge	HSE team members are responsible for the following: • Conducting prequalification verifications and evaluations of contractors added to the owner's approved contractor lists. • Supporting the SOW owner with contractor oversight activities and conformance assurance per this standard. • Facilitating field-level contractor safety engagements. • Ensuring tasks are risk assessed in alignment with the people safety standard. • Verifying contractor safety supervision is established and adequate for the SOW. • Leading, performing and/or supporting contractor audits and inspections when required. • Supporting the development of a safety improvement plan when required.
Frontline Leaders	Knowledge	Frontline leaders are responsible for: • Where applicable, performing reviews of the contractor's applicable documentation—practices, procedures, and supporting documents—for adequacy and conformance with the owner's requirements. • Understanding and applying oversight requirements as required for the given contractor or supplier. • Managing contractor performance at the site while the work is occurring. • Monitoring performance and intervening with peer management when required.
Workers	Awareness	All workers are responsible for: • Ensuring the owner's contractor and supplier procurement process is followed when selecting contractors and suppliers. • Ensuring that completed contract work is consistent with the owner's expectations, requirements, and documented SOW. • Informing owner leadership of any concerns regarding contractors and suppliers.

Requirements of This Standard

Prequalification and Selection

Contractor and Supplier Selection

Contractor and Supplier Procurement Process

The owner recognizes the importance of clear and effective procurement processes for contractors and suppliers in supporting its operations. Therefore, the owner shall ensure a robust process is in place to govern the acquisition of goods and services. This process shall include, at a minimum:

- **Clear procurement guidelines:** Establishing detailed guidelines that outline the procurement strategies to be employed by specific disciplines, based on contract value thresholds.
- **Procurement strategy requirements:** Including specific requirements for competitive sourcing, negotiated solution sourcing, and sole sourcing of contractors and suppliers.
- **Approval hierarchies and thresholds:** Defining clear approval hierarchies and spend thresholds for different levels of procurement activities, ensuring appropriate oversight and accountability.
- **Documentation and reporting standards:** Establishing documentation practices to record all procurement activities, ensuring transparency and accountability in decision-making processes.
- **Training and awareness programs:** Providing ongoing training and awareness programs for employees involved in procurement, emphasizing ethical practices, regulatory compliance, and risk management.

Contractor and Supplier Prequalification

Contractor Selection Criteria

Contractors seeking to work for an owner enter the owner's prequalification process to ensure the owner's requirements are met before the work can be awarded. The owner shall evaluate and select contractors based on criteria that include, among other dimensions, the following:

- Capability and competency
- Operational experience
- Financial viability, including insurance
- Current and historical health, safety, and environmental (HSE) performance
- Location/community
- Risks associated with the SOW
- Service quality—quality assurance/quality control (QA/QC)

Supplier Selection Criteria

The owner shall ensure processes are in place to evaluate and select suppliers based on criteria that include, among other dimensions, the following:

- Capability and competency
- Operational experience

- Financial viability, including insurance
- Location/community
- Service quality—quality assurance/quality control (QA/QC)

Contractor Expectations	The owner shall ensure a process is in place to communicate the owner's expectations and policies to contractors. This process shall include, at a minimum: • The owner's policies, including drug and alcohol policy, and respectful workplace policy • The owner's specific training and competency requirements • Expectations and requirements regarding subcontractors • Contractor safety expectations contained within the owner's safe work practices and other IMS practices or supporting documentation
Prequalification—Contractors and Critical Suppliers	The owner shall ensure a process is in place to prequalify contractors and critical suppliers in alignment with the owner's expectations, prior to being awarded work. The owner's HSE team shall administer the contractor and critical supplier prequalification process. • Prequalification shall include assessment of contractors and critical suppliers. • Following assessment, prequalified contractors and critical suppliers shall be added to the owner's approved contractors and suppliers list.
Approved Lists for Contractors and Suppliers	The owner shall ensure that only approved contractors and suppliers are engaged to work for or with the owner. Companies shall be identified on an approved contractor and approved supplier lists. The owner shall maintain two registers for companies that have been successfully prequalified to be engaged by the owner for work: • **Approved Contractor List:** A listing of contractors within the prequalification database approved to supply services and perform work on an owner worksite. • **Approved Suppliers List:** A listing of critical suppliers within each operations discipline from which buyers may procure materials, equipment, and machinery.

Contractor and Supplier Evaluation

Contractor Evaluation	After the prequalification process and prior to approval, the owner shall evaluate bidding contractors based on their HSE performance and technical (including quality management) capabilities. This evaluation contributes to the decision-making process for selecting the contractor that will be granted the work. The owner's process for evaluating and selecting contractors shall include the following considerations: • Ability to meet SOW and technical requirements • HSE Plan for the SOW, processes, and practices, as required

- HSE management systems/safety programs
- Contractor technical and HSE performance history
    - WCB experience and loss prevention history both on owner worksites and within the industry
    - Current and past enforcement actions and HSE charges against the contractor
- Compliance with regulatory and statutory standards
- Financial analysis (outlined below)
- Ability to execute work safely, reliably, and efficiently relative to the known hazards and risks associated with the SOW
- Weighted score allocations for evaluation criteria at a minimum shall consider—HSE, technical, quality, local content, and price

Supplier Evaluation

The owner shall ensure that processes are in place to evaluate all suppliers based on accessibility, reliability, quality (QA/QC), pricing, financial analysis, and other unique or specific criteria prior to use.

The owner shall ensure that financial verifications are performed and maintained on all critical suppliers in alignment with the owner's risk tolerance and financial management requirements.

Contractor and Supplier Financial Evaluation

The owner shall ensure that financial analysis of potential contractors and suppliers is conducted based on the anticipated contract's annual value as outlined below:

- For all contracts exceeding $5 million annually, financial analysis shall be performed by the treasury department.
- Contracts with an annual value between $1–5 million shall undergo financial analysis at the discretion of the approver.
- Contracts with an annual value up to $1 million do not require financial analysis, with decisions made by the project manager.

Financial analyses shall be conducted by the treasury department, encompassing reviews of vendor performance history, financial statements, and potential legal implications. Contracts exceeding $1 million annually shall involve treasury guidance in case of identified heightened financial risks during the contract period.

Competency—Confirmation of Contractor Competency

Training Requirements for Contractors

The owner shall ensure processes are in place to communicate the owner-specific training requirements to contractors and to verify that contractors working on location have completed the minimum contractor training requirements and relevant task-specific training.

Contractor Regulatory Obligations for Training

The owner's expectation is contractor compliance with legislation to provide their employees with necessary training to perform their assigned tasks safely, reliably, and efficiently.

When verified by the owner, training shall be:

- Relevant to the contractor employee's job
- Conducted by qualified instructors
- Contractors have verified as effective through employee testing and knowledge checks
- Records are documented and maintained within employee records

Contractor's Training and Competency Responsibilities	The owner shall maintain processes for ensuring contractors accept their responsibilities for:
	• Ensuring only competent workers are assigned to perform work and tasks on the owner's worksites.
	• Filling roles with qualified and competent workers who are fit for duty.
	• Providing adequate instructions, supervision and/or training when qualified workers are not available.
	• Verifying that training and competency meets the following:
	  ○ Regulatory requirements
	  ○ Industry-recommended practices
	  ○ Manufacturers' recommendations
	• Providing direct supervision of non-competent workers and implementing a program to develop and track worker task proficiency until the worker is fully competent in the assigned role.

Enhanced Controls for Critical Contractors and Suppliers

Controls for Contractors

Enhanced Controls for Critical Contractors	Contractors deemed essential to operations due to the nature of their work, whose involvement presents a heightened risk of safety incidents and schedule delays, are considered critical contractors. This designation requires additional communication and engagement to mitigate potential performance risks and ensure compliance with operational standards and expectations. The owner shall ensure enhanced controls are in place for critical contractors who are essential to operations and present increased performance risks due to the nature of their SOW. Enhanced controls shall include, at a minimum:

Controls	Details
Pre-Work	• In-person meetings to align on SOW • Contractor readiness verification, which shall include, at a minimum: ○ Validation of procedures needed to execute the SOW.

	◦ Validation that workers executing the tasks have competencies to execute the procedures and qualifications to operate equipment. ◦ Supervisory requirements and execution planning ◦ Verifications that the contractor is made aware of the general hazards and risks associated with the SOW and on the owner's worksite. • Project kick-off meetings
Ongoing-Work	• Regular owner and contractor leadership and HSE field visits • Regular monitoring of lagging indicators, management, and intervention on leading Indicators • Regular contractor HSE field assurance and performance management (schedule, quality, and cost) checks ◦ Safety improvement plans are highly recommended to address negative trends in performance ◦ Service quality incident investigations (as per the Incident, Emergency and Security Management Standard) • Quarterly meetings with leadership to discuss performance and opportunities for improvement/ realignment
Post-Work	• Annual or project lookbacks

Operational disciplines shall maintain an active list of critical contractors and ensure enhanced controls are implemented to ensure ongoing oversight and effective risk management.

Controls for Critical Suppliers

Enhanced Controls for Critical Suppliers

Suppliers deemed essential to operations due to the provision of materials, equipment, and machinery that are crucial to operations are considered critical suppliers. These suppliers may significantly impact safety, reliability, availability, schedule, and cost performance.

The owner shall ensure enhanced controls are in place for critical suppliers who are essential to operations.

Enhanced Controls for Critical Suppliers shall include, at a minimum:

Controls	Details
Pre-Order	• Perform prequalification, which may include onsite audit/verification to ensure critical suppliers align with the owner's expectations and standards for safety, reliability, and efficiency. • Confirm quality expectations, delivery schedules, support commitments, and penalties for non-compliance.

Upon Delivery	• Verify that all supplied materials, equipment, and machinery meet specified quality standards and industry regulations. Perform thorough inspections and testing upon receipt to confirm compliance with the owner's requirements. • Maintain records of all supplied materials, equipment, and machinery, including purchase orders, inspection reports, and maintenance logs.
Operational Use	• Regularly review supplier performance, including safety concerns, reliability, delivery times, defect rates, and support responsiveness. • Verify that suppliers provide comprehensive support, including installation, training, maintenance, and troubleshooting. • Regularly communicate through channels established to address issues and updates promptly.

Operational disciplines shall maintain an active list of critical suppliers and ensure enhanced controls are implemented to ensure ongoing oversight and effective risk management.

Performance Management

Leadership and Management Engagement

Contractor Engagement

The owner shall ensure processes are in place to ensure that contractors and suppliers align with owner specifications, with adherence to expectations for safe, reliable, and efficient operations. Leadership and management engagement shall:

- Be designed such that supervisors and managers from both the owner and contractors and/or suppliers have open and predictable communication paths that support achievement of safe, reliable, and efficient performance goals.
- Respective party leaders are engaged at frequencies deemed adequate for the work being executed.
- Include the following:
    - Executive-level relationships for long-term strategic contractors and/or suppliers
    - Manager-level relationships with contractors and/or suppliers
    - Site-level peer-to-peer routine engagements to enable tactical performance
    - Quarterly contractor safety meetings/forums that bring all contractor management and owner management together for collective engagement and business requirements alignment

Engagement Types

Frequent and regular management engagement is required to ensure that performance-related information is communicated to the contractor workforce.

Management engagements may include:

- Contract kick-off meetings, orientations, or onboarding sessions and premobilization
- Joint worksite visits and observations
- Contractor management safety teams
- Meetings—focused on contract deliverables and performance
- Safety stand-downs.

Monitoring and Evaluating Contractor Performance

Performance Management Criteria

The owner shall monitor and evaluate contractor and supplier performance on an ad hoc basis in an ongoing manner, communicating this performance to the contractor/supplier during the work—or at the conclusion of the work.

Monitoring and evaluating contractor SOW delivery performance includes:

- Developing and implementing performance criteria relevant to the contractor/supplier and the work being executed
- Acquiring the information through reporting, preferably using the owner's systems where practicable
- Analyzing performance data to identify trends and insights
- Soliciting performance feedback from internal stakeholders
- Coordinating periodic meetings held between the contractor and owner as required to discuss any performance improvement needs, including negative trends in KPIs
- Addressing and investigating service quality concerns or incidents as per the Incident, Emergency, and Security Management Standard

The owner shall ensure that processes are in place to address SOW performance concerns with the contractor. This process shall include mechanisms to limit liability, recover loss, and terminate agreements.

Contractor Performance Evaluation

Contractors' periodic and overall performance evaluations shall be the responsibility of the SOW owner (project manager).

- This exercise is undertaken with input from all stakeholder groups
- The final performance evaluation shall be completed on an annual basis or upon SOW completion

Verifying Compliance with the Owner's Requirements

Compliance Verification

Compliance verification of contractor activities is required to ensure that contractors are complying with the owner's requirements while working on the owner's worksites.

Compliance verification shall include the following facets:

- Compliance with IMS and HSE requirements

- Compliance with the contractors'/suppliers' own procedures and processes
    - Shall include field verification of use or spot checks/ inspections
- Contractors shall conduct and document assurance verification on their crew's work performance on owner worksites and communicate the results back to the owner.
- The owner shall periodically and on an ad hoc basis verify contractor compliance with its requirements.

Compliance verification can take many different forms and will be determined on a contractor-by-contractor basis. Activities shall include:

- Direct observation of work performed at the worksite through spot checks and inspections
- Audits at the contractor administration offices
- Submission of reports from the contractor to the owner

Contractor HSE Inspections and Audits

The owner shall ensure processes are in place to verify contractor performance through operations assurance activities, aligned with the Operations Assurance Standard.

Contractor inspections and audits will integrate management system alignment and HSE performance into the monitoring and evaluation of contractors.

- Contractor inspections and audits shall be conducted using approved templates, documented in the designated SOR, and communicated to the SOW owner (project manager), IMS administration team, and senior leadership.
- Deficiencies identified during inspections or audits shall be communicated to the contractor for prompt action and resolution.

Managing Safety Performance

Safety Improvement Plan (SIP)

A Safety Improvement Plan (SIP) is a structured and time-bound framework designed to address and enhance contractor performance that does not meet the owner's safety standards. SIP shall foster collaboration between the owner and the contractor to implement necessary changes effectively.

The owner shall ensure a process is in place to ensure improvement plans are developed and executed whenever contractor performance falls below acceptable thresholds. Specifically, an improvement plan is mandated for a contractor to continue working on the owner's sites under the following conditions:

- Failing to achieve an acceptable performance assessment score
- Demonstrating concerning trends related to the owner's life-saving rules inspections, audits, or other leading and lagging indicators

- Exhibiting concerning trends related to safety incidents, including near-miss events
- Receiving feedback from the owner's field representatives

The improvement plan process shall include, at a minimum, the following:

- Development of an approved improvement plan template which shall include, at a minimum:
  - SOW details: services provided, work location, start/end date.
  - Reason for issuing the improvement plan requirement: specifics of the criteria met.
  - Actions to address deficiencies: tracking and updating closure, deadlines, and responsible contractor personnel.
- Clear roles and responsibilities identifying:
  - The owner's SOW owner, in collaboration with the owner's HSE team, as the initiator of the improvement plan with the contractor.
- Retention of the improvement plan records in the owner's prequalification database.
- A mechanism to identify contractors working under the conditions of an improvement plan.
- A mechanism to terminate approved status should the contractor fail to address deficiencies or to improve performance to acceptable thresholds.

Managing SOW Performance

The owner shall ensure that processes are in place to ensure performance is consistent with established goals and targets throughout SOW execution.

Verification and Measures

Overview

Verifications and measures (V&Ms) are used to articulate the health of this standard and describe progress toward defined goals and targets. V&Ms are developed, established, and approved by the owner to:

- Confirm understanding of requirements by practitioners and leaders involved with executing the standard.
- Measure executed work and identify areas of deficient performance.
- Trend performance over time and illustrate the results of corrective actions.
- Support the continuous improvement, and therefore effectiveness, of the standard as a whole.

Selected measures shall be defined as key performance indicators (KPIs) that are used with higher levels of management as strategic indicators of performance.

Verification Activities

Verification activities may include the following:

- Leadership knowledge of contractor performance management principles and the basic contractor lifecycle model.

- Site leadership confirmation of onboarding and orientation, and knowledge of active work supervision requirements.
- Leadership awareness of common contractor performance controls, including audits and inspections and other management engagements.

Leading Measures

Leading Indicators may include the following:

Expectations	Purpose of Measure	Method of Measure
Accountability	IMS and HSE performance of contractors/suppliers is monitored.	• Number of proactive joint owner-contractor field visits performed by the owner.
Prequalification	Contractors engaged on owner worksites are prequalified before work is awarded.	• Spot checks prequalification Database for verification of use of prequalified contractors.
Risk Management	Risks associated with contractors are reduced due to ongoing engagement.	• Completion rate for quarterly meetings with applicable critical contractors.
Management Engagement	Owner/contractor leadership routinely engage to discuss HSE performance and collaboratively improve.	• Number of management-contractor engagement sessions completed. • 100% completion rate for quarterly engagements with applicable contractors.
Performance Management	Contractor audits and inspections are completed.	• Number of audits completed • Number of inspections completed. • Number of service quality investigations completed.

Lagging Measures

Lagging indicators may include the following:

Expectations	Purpose of Measure	Method of Measure
Prequalification	Contractors employed on the owner's worksites are prequalified before being awarded work.	• Number of events tied back to prequalification.

Risk Management	Risks associated with contractor work execution are identified and controlled.	• Number of events tied to risk assessment and management deficiencies.
Competency	Contract workers are competent for the roles performed.	• Number of events tied back to contractor competency deficiencies.
Performance Management	Contractor audits are completed.	• Number of performance management issues identified in contractor audits.

APPENDIX 2 CONTRACTOR HSE CONTENT TO BE REVIEWED

Date:		(DD/MM/YY)
Process:	Review HSE Manual to Determine if Covered for Each Contractor	
Project:		
Prepared by:		
Legend:	**X**	Absent from Contractor's HSE Processes
	√	Present in Contractor's HSE Processes

	Contractor				
Owner's HSE Requirements	**1**	**2**	**3**	**4**	**5**
Leadership Commitment and Accountability					
Contract HSE Plan					
HSE alignment and communication of expectations					
Performance recognition					
Safe Operations					
Site access requirements					
Fit for duty / fit for work					
Alcohol, drugs and substance abuse					
HSE violations					
Responsibility to refuse unsafe work					
Behavior-based safety – leading safety conversations					
Hazardous materials					
Ground disturbance					
Personal protective equipment (PPE)					
Hair regulations					
Noise / hearing requirements					
Fall protection					
Right of way and roadside work					
Power lines and high voltage work areas					
Safeguards and barricades – excavations, trenches and openings					
Radiation-producing equipment					
Mechanical equipment and motorized vehicles					
Contractor owned fabricated equipment					
Life-saving rules					

Risk Assessment and Management					
Hazard identification, risk mitigation and control					
Permit to work authorization					
Management of change					
Risk matrix					
Emergency Preparedness					
First aid and medical					
Emergency response plan					
Trained and competent emergency response team (ERT)					
Reliability And Integrity					
Personnel Competency and Training					
General requirements					
Short service employees					
Incident Management					
Reporting and investigation of incidents					
Communication with media					
Post-incident notification of external stakeholders' regulators and authorities					
Environmental Stewardship					
Management Of Change					
Information Documentation And Effective Communication					
Information and documentation					
HSE meetings - General safety meetings					
HSE meetings - Toolbox talks					
HSE meetings - End of shift reviews					
Compliance Assurance and Regulatory Advocacy					
Inspections					
Audits and evaluations					
Design Construction Commissioning, Operating and Decommissioning					

Contracted Services and Materials					
Performance Assessment and Continuous Improvements					
KPIs					
Tours					
Best practices and continuous improvements					

APPENDIX 3 SAMPLE CONTRACTOR HSE EVALUATION AND SCORING PROCESS

HSE Evaluation Completed By:		Date Completed:		Contractor:	

	HSE Evaluation Criteria	Evaluation					
		Show Stopper (SS) or Red Flag (RF)	Weight	Complete Y, N, or N/A	Quality 2, 1, 0	Score	Evaluator's Comments
1	Prequalification has been completed.	SS	3			-	
2	The contractor has read and understands the owner's contract HSE requirements.		1		-	-	
3.1	Preliminary contractor health and safety plan		5			-	
3.2	High level hazard assessment and control plan		4			-	
3.3	HSE procedures and practices for this scope of work have been identified.		4			-	
4	The company has an industry recognized certificate of performance and compliance	RF	3		-	-	
5	Was the HSE MS or equivalent audited by an external auditor? If yes, provide details of the HSE audit (date, auditor, protocol, score).	RF	2		-	-	
6	On completion of the HSE audit, was a written action plan developed to address the audit findings and correct the deficiencies identified?		3			-	
7	Was the HSE MS or equivalent audited by an internal auditor? If yes, provide details for HSE audit (date, auditor, protocol, score).	RF	2		-	-	
8	On completion of the HSE audit, was a written action plan developed to address the audit findings and correct the deficiencies identified?		2			-	
9	Workers compensation board information for each of the last three years (employer and industry).	SS	3		**0**	-	
	(columns: This Year / Last Year / 2 years ago)			-	-	-	
	Employer premium rate			-	-	-	
	Industry rate			-	-	-	
	Rate adjustment %			-	-	-	
	Surcharge or discount			-	-	-	

HSE Evaluation Completed By:		Date Completed:		Contractor:	

	HSE Evaluation Criteria										Evaluation					
									Show Stopper (SS) or Red Flag (RF)		Weight	Complete Y, N, or N/A		Quality 2, 1, 0	Score	Evaluator's Comments
10	Safety Performance	Current Year			Last Year			2 Years Ago			5				-	
	E-Employee / C- Contractor / T-Total	E	C	T	E	C	T	E	C	T			-	-	-	
	Exposure hours worked			0			0			0			-	-	-	
	Fatalities*			0			0			0	SS	-3	Y		-	
	Lost time injuries (LTI)*			0			0			0		-2	Y		-	
	Restricted work case (RWC) injuries			0			0			0			-	-	-	
	Medical treatment (MT) injuries			0			0			0			-	-	-	
	First aid (FA) injuries			0			0			0			-	-	-	
	Motor vehicle incidents			0			0			0			-	-	-	
	Ground disturbance incidents			0			0			0			-	-	-	
	Total number of recordable injuries	0	0	0	0	0	0	0	0	0			-	-	-	
	Total recordable injury rate (TRIR)	-	-	-	-	-	-	-	-	-	RF	1	Y	-	**1**	
	Lost time injury rate (LTIR)	-	-	-	-	-	-	-	-	-	RF	2	Y	-	**2**	
	Total injury frequency (TIF)	-	-	-	-	-	-	-	-	-	RF	1	Y	-	**1**	
11	Was the company cited, charged or prosecuted for any occupational health and safety non-compliance or environmental offence in the past three (3) years?								RF		-2				-	

HSE Evaluation Completed By:		**Date Completed:**		**Contractor:**			
	HSE Evaluation Criteria	**Show Stopper (SS) or Red Flag (RF)**	**Weight**	**Complete Y, N, or N/A**	**Quality 2, 1, 0**	**Score**	**Evaluator's Comments**
12	Was the company issued with a stop work order from/by a government regulatory agency in the past three (3) years?		-1			-	
13	Were any employee from the company barred from working on any owner site due to a HSE violation?		-1			-	
14	Has a regulator ever had cause to investigate a company worksite in the past three (3) years?		-1			-	
15	Has the company had any reportable / notifiable incidents to regulators (safety, occupational health and environmental) in the past three (3) years?		-1			-	
16	Has the company been involved in any reportable spills or releases (reported to regulators) in the past three (3) years?		-1			-	
17-18	Evidence (actual samples of completed records) of the subcontractor management process in use for: 1. Selection / evaluation of subcontractors. 2. Monitoring subcontractor performance.	SS	2			-	
19	Has the company provided a copy of its HSE Policy	SS	4			-	
20.a	Has the company provided the following documentation: • A copy of its HSE management system manual (including safe work practices associated with the scope of work), which align with the owner's requirements	SS	5			-	
20.b	• Fall protection (if work involves working from heights)	SS	4			-	
20.c	• Lock out / tag out energy isolation	SS	4			-	
20.d	• Alcohol and drug testing policy	SS	4			-	
20.e	• Ground disturbance (if work scope involves ground disturbance)	SS	4			-	
20.f	• Confined space entry (if work scope involves confined space entry)	SS	4			-	
20.g	• Disciplinary process to addresses fighting, possessing a dangerous weapon or making false statements	SS	3			-	
20.h	• Facial hair	SS	2			-	
20.i	• Smoking only in designated areas	SS	2			-	
20.j	• Field level hazard assessment (FLHA) program	RF	3			-	
20.k	• Vehicle driving and cell phone use policy	RF	2			-	
20.l	• Defensive driving training	RF	2			-	
20.m	• Owner approved PPE usage	RF	3			-	
20.n	• Management of change (MOC) process	RF	2			-	

HSE Evaluation Completed By:		**Date Completed:**		**Contractor:**	

	HSE Evaluation Criteria	**Evaluation**					
		Show Stopper (SS) or Red Flag (RF)	**Weight**	**Complete Y, N, or N/A**	**Quality 2, 1, 0**	**Score**	**Evaluator's Comments**
20.o	• Environmental stewardship program	RF	2			-	
20.p	• Progressive discipline program	RF	4			-	
20.q	• Safety meetings	RF	2			-	
20.r	• Learning from incidents built into the incident management process	RF	4			-	
21	Has the company provided a copy of its emergency response plan and/or procedure	SS	2			-	
22	Has the company provided a completed incident report / investigation / lessons learned	SS	4			-	
23	Has the company provided evidence (actual samples of completed records) of a follow-up action process in use (e.g., associated with incidents, audits, inspections)	RF	4			-	
24	Has the company provided its current HSE objectives / goals / targets for the current year		2			-	
25	Has the company provided a copy of the organization chart (with names) for personnel supporting the contract SOW.		2			-	
26	Has the company provided specific orientation / training for supervision with respect to their HSE responsibilities and accountabilities. Evidence (actual samples of completed records) of content and completion records.	RF	4			-	
27	Has the company provided evidence of a supervisor development program	RF	4			-	
28	Has the company provided resumes for HSE personnel supporting the contract SOW		2			-	
29	Has the company provided evidence of a development plan for HSE personnel supporting the contract SOW		4			-	
30	Has the company provided copies of job description(s) for the positions that provide services as per the SOW, including training and competency requirements for each position		4			-	
31	Has the company provided a copy of its training policy or procedure and competency assurance process (should include a training matrix).	SS	2			-	
32	Has the company provided evidence of training record(s) and competency assurance record(s) (actual samples of completed records)	SS	2			-	
33	Has the company provided evidence of new hire orientation (actual samples of completed records), orientation content and quiz / checklist	RF	2			-	
34	Has the company provided evidence of New worker to site / green hat program	RF	3			-	

HSE Evaluation Completed By:		**Date Completed:**		**Contractor:**			
HSE Evaluation Criteria		**Evaluation**					
		Show Stopper (SS) or Red Flag (RF)	**Weight**	**Complete Y, N, or N/A**	**Quality 2, 1, 0**	**Score**	**Evaluator's Comments**
35	Has the company provided evidence (actual samples of completed records) of your return-to-work process (modified work program, medical accommodation).		2			-	
36	Has the company provided a copy of evidence (actual samples of completed records) of background verification and fit for duty assessments (pre-hire).	RF	2			-	
37	Has the company provided a copy of evidence (actual samples of completed records) of workplace inspection processes	RF	2			-	
38	Has the company provided a copy of evidence (actual samples of completed records) of equipment certification and inspection processes.	RF	2			-	
39.a	Has the company agreed to the owner performing a HSE assessment on its organization at an active project site? If so: • Is a list of active project sites that the owner could potentially visit to complete an HSE assessment and HSE contact name(s) been provided.		2			-	
39.b	• Rank feedback from the active project site HSE references		5			-	
40	List of owner's sites/projects where work is/was completed. Provide owner HSE contact name(s). Rank feedback from the owner's references.		5			-	
41	Owner's HSE evaluator's overall satisfaction with contractor submission (HSE qualitative question)		5	Y		**0**	
42	Owner's HSE evaluator's overall judgement of contractor (HSE qualitative question)		5	Y		**0**	

Note: Much of this data from line 10–42 can be recorded in an effective prequalification database and can be reviewed within the prequalification database.

APPENDIX 4 PREBID MEETING FORM

Contract/Tender Number:	
Date:	
Work Scope Description:	
SOW / Project Owner:	
Completed By:	
Contractor Companies Present:	

The HSE Pre-Bid Meeting Form shall be reviewed during the pre-bid meeting with all contractors who are potentially going to perform "hands on" work activities. Additional items may be added to the form as required. The intent of the pre-bid meeting is to ensure that contractors are aware of owner HSE requirements prior to bidding.

	DISCUSSION ITEM	Applicable		Comments
		Yes	No	
1	Introductions – Purpose of Meeting / Safety Moment.			
2	Contractor alignment to the owner's commitment to HSE and operational excellence.			
3	Environmental Due Diligence: • Environmental Management Plan. • MSDS. • Hazardous Materials Brought on Site. • Spill Reporting. • Risk Management.			
4	Industrial Hygiene Resources / Monitoring.			
5	Contractor Management Process • Has the prime contractor for this project been established?			
6	Is the contractor prequalified in the owner's database • If not, enable the process and support as required.			
7	Subcontractor Prequalification and Management Process.			

	DISCUSSION ITEM	Applicable		Comments
		Yes	No	
8	Contractor compliance with Owner's HSE Requirements: • Experience of workforce / English literacy – communication challenges. • Drug and alcohol policy. • Smoking policy. • Modified work / medical accommodations program. • Contractor compliance mechanism for own HSE management system. • A thorough work scope review has been completed. • A walk through the work location has been completed. • Area / Job Specific Standards / Work Practices / Procedures / Rules / Owner reviewed. • Statutory Requirements / Owner / Employers / Worker / Competent Worker / Obligations reviewed. • Reinforcement that all workers adhere to Owner's fall protection requirements. • Requirements for an Owner Approved Contractor Project Specific Hazard Assessment and Control Plan be developed prior to Pre-Mobilization Meeting. Document must be submitted to Owner HSE in advance of the Pre-Mobilization Meeting. • Obligation to advise the Owner's Project Executions Project Executions Projects business area management of any unique hazards presented by the contractor's work, or any hazards found by the contractor during execution of the work. • Hazards assessment and control plan will be reviewed with employees performing the tasks. Ensure each contract employee is instructed in the known potential fire, explosion, or toxic release hazards related to his or her job and the process. • Quality Management Plans Requirements (Building / Gas / Electrical / Fire / Radiation / Elevators / ABSA) – as applicable.			
9	Material / equipment inspection and certification program is in place.			
10	The contractor will provide the owner with a list of all equipment to be used on site.			
11	Contractor job-specific organizational structure: names / contact numbers / e-mails / resumes / development plans.			
12	Commissioning – commissioning is the sole responsibility of owner operations unless otherwise agreed.			
13	The contractor will comply with the owner's site operating rules.			
14	The contractor will participate in the following: • Scheduled progress review meetings. • Kick-off meetings. • Pre-mobilization meetings. • HSE alignment workshops. • Audits and inspections.			

	DISCUSSION ITEM	Applicable		Comments
		Yes	No	
15	Have a risk management process and plan			

16	Have a training management process and plan
	• Ensure the use of only trained and competent workers.
	• Ensure each contract and subcontract employee has the appropriate qualification and job skill training to complete the contracted work.
	• Ensure each contract employee receives and understands training regarding the following:
	    ○ Owner's project execution safety rules and regulations and applicable HSE requirements.
	    ○ Applicable provisions of the emergency response and control plan.
	    ○ Applicable safe work practices and work permit requirements of the facility.
	    ○ Field level risk assessment (FLRA) / Field level hazard assessment (FLHA).
	    ○ Orientations.
	    ○ All on-site employees have completed the owner's site-specific and required training or accepted related industry equivalent – as applicable to province, state or country.
	• Maintain records to demonstrate that each contract employee has received and understood the instruction and training described above. Training records shall include the following:
	    ○ Identity of the contract employee including applicable ID numbers.
	    ○ Date of training.
	    ○ The material covered.
	    ○ The means used to verify the employee. Understood the training.
	    ○ Resume of trainers.
	• Provide the owner with a training matrix that identifies the applicable job skill and safety training for each contract employee at the job site. Periodically, actual training/ certification records will also be provided upon request.			

17	Have an emergency response plan.			
18	Have an HSE monitoring and performance management plan.			
19	Have a process for permit-to-work authorization.			
20	Have relevant safe work procedures for the SOW.			
21	Orientation for all workers prior to working on the site.			

	DISCUSSION ITEM	Applicable		Comments
		Yes	No	
22	Site access requirements • Site orientation. • Required PPE. • Government issued ID. • Pre-access alcohol and drug testing. • Fit for duty verifications.			
23	Be knowledgeable of the proper procedures and responsibility for the reporting of incidents and injuries to owner management.			
24	Other			

APPENDIX 5 ASSIGNING OWNERSHIP OF WORKSITE—PRIME CONTRACTOR TRANSFER

Subject:	Designation of Prime Contractor Status to Contractor	Date (Y-M-D)	yyyy-mm-dd
From:	Senior Owner Authority	Department:	As Applicable
To:	Contractor Senior Authority	Cc:	As Applicable

Prime Contractor status has been transferred from Owner	(Insert Owner Name)
To (Insert Contractor Name) for the duration of the SOW, effective	Insert Date Here

Requirements

According to the Owner's Project Execution Contractor HSE Management Requirements:

- The project manager shall ensure the following items are completed:
    - Prime contractors are prequalified and evaluated by owner HSE.
    - The owner has approved the prime contractor's HSE processes, including, but not limited to:
        - Process for controlling site access to personnel.
        - Emergency response plan.
        - Incident notification process (including injuries & fatalities, near misses, environmental, reputation, asset damage).
        - Regulatory inspection/visitation/requests notification process.
    - The owner shall issue a general permit to the prime contractor.
    - The owner shall issue a legal description including a map/plot plan with coordinates demarcating the area approved for the prime contractor.
    - A signed legal agreement along with the terms and conditions between the owner and the contractor has been approved by the owner's senior authority.
- Documents are files and retained with assigned contracts with the SCM/procurement organization.
- The owner supply chain manager and HSE manager shall verify that the above requirements have been fulfilled.
- Prime contractor designation shall be approved by the owner's senior leader.

Please refer to the attached information:

(Attach all relevant supporting information for communicating the transfer of prime contractor status to the contractor).

- Legal description including a map/plot plan with coordinates demarcating the area approved for the prime contractor.
- General permit to the prime contractor.
- Legal description including a site map with coordinates demarcating the area approved for the prime contractor.
- Process for controlling site access to personnel.
- Incident notification process.
- Emergency response plan.

Acknowledgment

By signing this form, all parties agree:

- (Insert Contractor Name) ____________________ will operate in the above-mentioned areas as the prime contractor.
- The requirements listed above have been completed.
- The contractor is aware of the hazards associated with the scope of work.
- The contractor is aware of and shall comply with the owner HSE requirements associated with the scope of work as the prime contractor.
- Any subcontractors used for this scope of work by the prime contractor shall comply with the prime contractors HSE requirements.

Function	Print Name	Signature	Date
Contractor Representative			
Owner Project Manager			
Owner SCM Manager			
Owner HSE Manager			

Approval

By signing this form, I confirm the requirements listed above have been completed.

Function	Print Name	Signature	Date
Owner Project Execution			

APPENDIX 6 WAIVER OF CONTRACTOR PREQUALIFICATION REQUIREMENTS

Requirements

1. This waiver shall be used only when there is no other prequalified HSE compliant contractor available to perform the required work.
2. The requestor shall ensure an HSE approved risk mitigation plan is attached to the waiver form.
3. The requestor shall ensure full oversight and supervision of the contractor during work execution while the contractor is not prequalified.
4. A waiver is valid for a specific scope of work (SOW), up to a maximum of 6 months. If the work must exceed 6 months, a new waiver shall be completed to ensure risk mitigation plans are adequate and effective, and to verify that use of a waiver is the appropriate business process for the SOW.
5. A waiver shall be approved by the owner, who shall be made aware of all associated risks assumed by the owner during the use of this non-prequalified contractor.
6. Completed waiver forms shall be forwarded to the HSE administrator to be recorded in the HSE contractor registry.

The following information must be completed for the use of non-HSE prequalified contractors or subcontractors on any owner worksite.

Contract/Tender Number:	
Contractor:	
Date:	

Owner—Requestor

I, ____________________ (Name), representative for ______________________ (Business Area) understand that ______________________ (Subcontractor or Contractor Name) will be utilized to perform work on our site and is not prequalified to do so according to the owner-contractor HSE management process. During the execution of this work, I accept full responsibilities for all activities and will ensure direct supervision for the duration of the work, and/or will ensure the contractor provides such direct supervision to its subcontractor (if applicable).

Scope of work:	
Reason for use of non-prequalified contractor:	

Summary of risk mitigation plan:	Refer to attached risk mitigation plan
Additional comments:	
Signature (requestor):	
Date:	

Contractor

I, ______________________ (Name), representative for ____________________ (Contractor Company) commit that during the execution of the proposed scope of work, while working on an owner work site (select one option):

- ¤ (Insert Contractor Name):
  - Will follow ______________________ (Contractor Company) policies and work practices.
  - Will follow owner health, safety & environment (HSE) policies and work practices, as well as the risk mitigation plan associated with this emergency waiver form.
  - Acknowledges they are not qualified to work at any other owner work site via this waiver form.
  - Additional requirements:______________________________
- ¤ (Insert Subcontractor Name):
  - Will follow __________________ (Contractor Company) policies and work practices,
  - Will follow owner health, safety & environment policies (HSE) and work practices, as well as the risk mitigation plan associated with this emergency waiver form.
  - As well as the risk mitigation plan associated with this emergency waiver form.
  - Acknowledges they are not qualified to work at any other owner work site via this waiver form.
  - Additional requirements: ______________________________

Acknowledgment

By signing this form, all parties agree:

- HSE topics listed above have been discussed.
- The contractor is aware of the hazards associated with the scope of work.
- The contractor is aware of the owner HSE requirements associated with the scope of work.
- To comply with all owner HSE requirements.
- That the contractor shall comply with the HSE mitigation plan referenced in this waiver.
- That any subcontractors used for this scope of work shall comply with the HSE mitigation plan referenced in this waiver.

Function	Print Name	Signature	Date
Contractor Representative			
Subcontractor Representative (if applicable)			
Contractor HSE Representative			
Owner Project Manager			
Owner Construction Manager			
Owner SCM / Procurement Representative			
Owner HSE Representative			

Approval: I acknowledge and approve the use of this contractor.

Function	Print Name	Signature	Approval Date
Owner Leadership Authority			

Expiration Date: ______________________________

APPENDIX 7 TEMPLATE FOR CONTRACT HSE PLAN

COPYRIGHT © Owner … *All rights reserved.*

Guidance to Contractors— The contractor may use this document (as a template or create a similarly structured document) to develop its contract HSE plan for the RFP (bid response) and for addressing the scope of work (SOW) hazards and associated risks. The plan shall describe how the contractor will manage HSE during the execution of the work. This document provides general guidance to the contractor. When completing the HSE plan, the contractor shall address all relevant requirements associated with the SOW.

Document Approval

Document Type	Plan		
Document Title	Template for Contract HSE Plan		
Document ID	Doc #		
Rev No/Date			
Role	**Name and Designation**	**Signature**	**Date**
Document Prepared By			
Document Prepared By			
Document Reviewed By			
Document Approved By			

Next Review Date:
Comment: Document review frequency might vary in case deemed necessary at the discretion of the approving authority document owner.

Purpose

Purpose	The purpose of this template and guidance included is to ensure the contractor provides the required HSE content within their bid response to ensure fair and equitable comparison and evaluation of bid responses of contractors. This HSE Plan template provides requirements that must be detailed in the contract HSE Plan. HSE requirements in this template are not exhaustive, and the contractor shall include relevant details as applicable. The contractor should select from this template those sections which are most appropriate to their response when addressing the SOW requirements and can add further sections as appropriate.

Scope

Scope

This template is recommended for use in all major contracts that require the contractor to provide an HSE Plan supporting the bid where an HSE requirement is defined in the contract SOW.

Responsibilities

Contractor Responsibilities

The contractor shall develop a preliminary contract HSE Plan as part of its bid submission. After award of the contract, the successful contractor shall update and implement a contract HSE Plan.

- The owner's HSE assigned leader and project manager are responsible for evaluating the adequacy of the preliminary contract HSE Plan of each contractor submitting a bid for the SOW.
- The HSE leader is responsible for working with the selected contractor to ensure that a complete and workable contract HSE Plan is developed, reviewed, and implemented.

Guidelines for the Content of Contract HSE Plan

Content of HSE Plan

The project manager should use this template to define the scope of the contract HSE Plan. The scope and contents of the contract HSE Plan should be specific to each contract and should reflect the nature of the work and the specific hazards and risks presented by the work.

- The project manager should also use this template when evaluating the adequacy of a contract HSE Plan.
- Contractors shall use this template to develop their preliminary and final contract HSE Plan when successful in securing the contract.

Depending on the nature of the work and at the discretion of the project manager, the contractor may provide some items in the templates at a lower level of detail or omit them completely at the bid stage. However, in doing this, the project manager should fully address the requirement elsewhere or provide a reason justifying the omission during the bid stage for the particular contract.

The assigned HSE representative for the contract bid reviews shall review the contractor's contract HSE Plan for completeness, and this shall be used in evaluating the contractor during the bid evaluation stage of the process.

The contract HSE Plan is the document which details how the contractor will manage HSE throughout the contract period. While executing the contract, use the contract HSE Plan document to monitor the contractor HSE performance. The guidelines assist both the contractor and the owner in the final preparation and execution of the contract (SOW) HSE Plans.

Description of a Contract HSE Plan

Description—HSE Plan

The contractor will develop the contract HSE Plan to define policies, programs, standards and procedures, specific actions that will be taken, and specific programs and procedures the contractor will implement to manage the hazards and risks associated with the SOW. This should cover all contract phases from mobilization through demobilization.

As a basis for developing its contract HSE Plan, the contractor should use the owner contract HSE documents where it may be included in the HSE specification and bid package documents of the contract.

However, the contractor should state its own proposals within the framework of these guidelines. In particular, the contractor should:

- Develop its own hazard and risks identification, assessment, controls, and recovery measures for the work described in the contract.
- Demonstrate familiarity with the type of work involved and the hazards and risks anticipated.
- Define its execution proposals (the contractor should confirm its proposed use of policies, procedures, and standards and show that these comply with the owner's requirements or add to them where it believes the contract guidelines are insufficient).
- Confirm its understanding that the owner's requirements represent minimum standards.

The contract HSE Plan is a living working document which may require updating as the contract progresses to rectify identified failings from audits/ inspections and reflect changes in the work scope.

Contract Health, Safety and Environment (HSE) Plan

Contractor HSE Plan

This contract HSE Plan should describe how the contractor will apply its established programs to manage its activities on the specific contract. The contractor should identify the contract-specific work activities it will perform, and the anticipated hazards, risks, and environmental aspects associated with the work. The contractor should clearly describe its plan for minimizing the potential consequences of these anticipated hazards, risks, and environmental aspects.

Such contract HSE Plans are to be commensurate with the work and in conformance with the specific requirements of the owner's major projects construction HSE reference manual, including any revisions thereto, provide the following content outlined in this section.

Contractor Leadership and Accountability

Leadership

The contractor leadership shall:

- Provide a statement of commitment to adhere to owner HSE requirements while working at site.

- Provide a list of the contractor senior management and a schedule for their participation in HSE activities related to the contract (meetings, inspections, self-assessments, etc.).
- Provide the contractor's written policies on HSE.
- Provide a schedule of meetings or other activities of line management participation in the communication and implementation of contractor HSE policy related to the contract.
- Provide procedures or safe work practices to cover life-saving rules (LSRs), but not limited to the following:
    - Working with the work permit system
    - Drug and alcohol policy
    - Verification of hazardous energy before work begins
    - Overriding safety-critical equipment without authorization
    - Fall protection
    - Working near high voltage equipment
    - Ground disturbance
    - Entering a confined space
    - Use of PPE
    - Safe driving—rules of the road
- A description of the disciplinary measures enforced resulting from non-compliance to HSE rules/regulations (if any).

Hazards Identification and Risk Management

Overview

The contractor shall:

- Provide a hazard identification and risk assessment plan using owner-identified hazards and risks and environmental aspects as a starting point, and any additional hazards and environmental aspects identified by the contractor.
- Identify control measures to mitigate the hazards and risks and manage the environmental aspects, including procedures and safe work practices.
- Provide a method statement to review the hazards and risks with the employees prior to performing the work.
- Provide a preliminary job hazard analysis for all high and medium-risks tasks inherent in the work scope.

Field Level Hazard Assessment (FLHA)

The contractor shall:

- Provide a procedure to conduct the FLHA prior to the start of each task, and/or when conditions change, or the workers have been away from the work area.

Contractor Environmental Management Plan (EMP)

The contractor shall:

- Describe how the contractor plans to apply its established, generic programs and how they will customize their plans to

manage the environmental aspects inherent in the specific contract.
- The contractor should identify contract-specific work activities they will perform.
- Explain how it anticipates managing and mitigating environmental impacts associated with the work. The contractor should clearly describe their plan for minimizing the potential environmental impacts of the anticipated aspects.

Attach the list of significant environmental aspects for this project to the contract HSE Plan. Refer to the list of significant environmental aspects below. If they identify additional significant environmental aspects, the contractor must include these in their EMP and inform owner. At a minimum, the significant environmental aspects/issues and potential impacts include, but are not limited to, the following environmental aspects listed below.

Environmental Aspects	Issues to be Addressed in the Contractor's Environmental Management Plan
Air Emissions	• Asbestos use and handling • Emission of smoke from plant and equipment • Open burning on site • Dust emission from construction and demolition activities • Idling of vehicles and equipment
Water Management	• Discharge of water—TSS, debris, chemicals • Concrete batching and pre-cast concrete casting wastewater • Hydro testing fluids • Water use • Groundwater—release, protection • Storm water and snow melt • Equipment wash water • Surface water protection • Any contaminated fluids
Noise	• General noise management • Powered mechanical equipment from general construction works • Control of piling noise • All vehicles, plants, and equipment (for noise)
Waste Management	• General waste management • Construction and demolition waste • Surplus excavated materials • Chemical waste (e.g., spent lube, sludge from oil interceptors, contaminated materials) • Asbestos waste • Recycling

Material Management	• Containers • Bulk storage and transportation, handling, and spill prevention
Spill prevention and containment	• Clean-up of spills • Operation and maintenance of equipment • Handling and spill prevention actions • Training of workers
Soil	• Sediment and erosion control • Soil salvage • Project soil management plan • Removal of soil—clay, sand, gravel, etc.
Control Products	• Approved list • Safety data sheet (SDS) • Handling and storage of chemicals and dangerous goods
Flora, Fauna and Historical Heritage	• Aquatic habitats • Vegetation and trees—clearing or disturbance • Fish and amphibians • Birds nest interruption • Wild animals • Historical heritage—discovery of sites, objects, etc.
Resource use	• Fuel and diesel • Electricity • Lubricant oil • Wood for formwork and trench support
Response to Environmental Emergencies	• Water site runoff during or after rainstorms • Chemical spillage • Discovery of historical sites • Discovery of rare/protected species of flora and fauna

- Contractor environmental management plans: describe the environmental control procedures and detailed mitigation measures to minimize the impacts of environmental aspects identified.
- Reference the controls suggested in the generic project's hazard management plan.
    - If there is an identification of additional significant environmental aspects, the contractor must include these in their EMP and inform the owner to update the list of significant environmental aspects.
- The environmental management procedures and mitigation measures must contain detailed information regarding the permanent and emergency environmental management systems, including, but not limited to the following:

  - The number, location, size and type of environmental management and treatment facilities, equipment, stations, containers, and kits
  - Details of chemical usage and disposal methods
  - Operation, monitor and test plans, inspection, maintenance, self-assessment plan, and procedures
  - Provide the sources of any water usage and details of the treatment and disposal of the wastewater
  - Information regarding the scheduling and sequencing of construction environmental control facilities prior to construction activities
  - Provide specific environmental report details
  - Details on subcontractor plans for any activities and details of incorporating their work into the contractor environmental control system
  - Provide details on how to incorporate environmental aspects into JHAs and FLHAs (i.e., checklists)
- **Waste Management:** Insert an owner waste management policy in this section. Otherwise, if necessary, provide a specific waste management policy for this project.
- **Classification and Estimation of Construction Materials:** Insert relevant types of construction materials and, in addition to their estimated quantities, throughout the period of project. Apply an example waste flow table (WFT) to this section.

Table A7.1: Waste Flow Table (WFT)

Type of Construction Material	Estimated QTY (m^3)	Duration of Material Generation	Remark
Construction waste to be delivered toowner disposal bins.			
Construction waste to be recycled			
Construction waste to be reused			

- Provide a description of how the contractor will maintain records of quantities/types of all wastes and provide plans for restoration of site to pre-work condition. This requirement is dependent on the project-specific waste management plan in place.
- **Measures to Reduce/Minimize Waste Material Generation:** The contractor must list work processes or activities that will generate construction waste materials. Provide the proposed measures to reduce/minimize such generation—in particular, metallic waste, timber, paper/cardboard packaging, and chemical waste.
  - General disposal
  - Reuse
  - Recycling

- Provide targets to reduce materials known to harm the environment.
- Provide a system of work for on-site sorting of construction materials. The system will include the identification of the source of generation, estimated quantity, and arrangement for on-site sorting.
- The owner may arrange for collection, temporary storage facilities, frequency of collection by recycling contractor, or frequency of removal off-site for:
    - Excavated materials
    - Construction and demolition materials
    - Chemical waste
    - General waste
- Depending on the project, the owner will arrange with the potential contractor managing the recycling program to collect materials from the site and record the quantities before removal. All contractors will, however, be responsible for source separation. The project environmental advisor of the project should be contacted for guidance.
- Provide details of a regular inspection program to monitor the implementation of this environmental management plan.

WHMIS Communication Program

The contractor's WHMIS communication program shall:

- Identify a list of all hazardous materials that will be brought to site and review owner safety data sheet (SDS) database to confirm approval status.
- Provide a list of all hazardous materials.
- Demonstrate the availability and distribution of guidance/ information on the SDS of chemicals likely to encounter in the contract and proposals for confirming adherence to the SDS during the contract.
- Demonstrate that the contractor has received, or provide a plan on how to receive, the required Industrial Hygiene/Environmental Affairs approval for these products.
- Identify WHMIS Communication Program responsibility and accountability.
- Demonstrate receipt of SDSs for materials to bring onto the jobsite by contractor or its suppliers and Subcontractor of any tier.
- Provide details of employee training requirements on SDS and in the handling and disposal of materials that fall under the owner's procedures and/or statutory regulations.
- Provide a disposal plan for removal of hazardous materials from the Jobsite. This plan must meet the requirements of the owner, and all federal, provincial, and other applicable governmental requirements.
- Notwithstanding any of the established owner or contractor programs as required, the Contractor is responsible for maintaining proper safety, fire prevention, hazard communication, and security conditions at the Jobsite.

Legal Requirements and Commitment

Leadership and Accountability	Legal Requirements and Commitment is a joint owner and contractor responsibility for overarching legislation. It is the owner's responsibility to obtain the permits, and the contractor's responsibility to review and follow the permits. The contractor will inform the owner to update the legal register when identifying additional environmental legislation to update the legal register.
	
	Legal Requirements and Commitment shall:
	
	• Provide a list including:
	  ○ Any applicable legislation governing the work scope.
	  ○ Contractor-identified regulations, codes, and standards applicable to the standard.
	  ○ Specific regulatory permits required to complete the job.
	• Provide details of statutory requirements/owner/employers/workers/competent work obligations.
	• Provide details of the contractor environmental due diligence requirements.
	• Provide a statement of how the contractor will ensure compliance with all applicable safety and health-related laws and regulations and directives of governmental and other agencies of competent jurisdiction.
	• Regulatory Permits and Programs: The contractor shall initiate and maintain such permits and programs as may be necessary to comply with requirements set forth by the regulator and any other local, provincial, territorial, and federal regulations.
	• The owner will provide a copy of all permits prior to commencement of work at the jobsite.
	  ○ If permits are not required to perform work, the contractor should submit a letter to the owner prior to commencement of work at the jobsite stating that no permits are required.

Goals, Targets, and Planning

Goals, Targets, and Planning	A plan to include gathering data and use of reactive performance indicators: • Recordable Injury Frequency (RIF)/Total Injury Frequency (TIF) • Restricted Work • Equipment Damages • Vehicle Incidents • Environmental Spills Key performance indicators (KPIs) must be available and use of performance records, to provide feedback/review/discussion at HSE meetings.

Management of Change

Management of Change	Refer to the Project Management section of the contract.

Structure, Responsibility, and Authority

Structure, Responsibility, and Authority	Structure, Responsibility, and Authority shall:
	• Provide a job-specific organization chart
	• Define line management, employee, and subcontractor responsibilities for environmental protection
	• Describe roles and environmental responsibilities of project key personnel. The responsibilities to include all aspects of operations, monitoring, testing, reporting, maintaining, and other activities associated with the planned environmental management system
	• Provide a list showing personnel (names, phone numbers, and email addresses of key contacts) responsible for the implementation of the contract
	• Provide job descriptions showing each team member's HSE competencies, responsibilities, and function
	• Provide job descriptions of the contractor's HSE professionals
	• Provide a statement/plan showing how to communicate HSE issues between the owner, contractor, and subcontractor (identify interfaces for all phases)
	• Provide a statement/plan to ensure that all employees are English literate and are 18 years or older to work on the job site
	• Identify the individuals charged with HSE responsibility at middle/senior management level, and their reporting structure
	• Indicate the reporting relationship of the contractor HSE advisers with line management and provide evidence of how line management follows advice offered
	• Provide a statement indicating the involvement of the HSE department in:
	    ○ Formulation and suitability of HSE rules
	    ○ Specific review and approval of all work plans and methods
	    ○ Inspections and audits
	    ○ Training
	    ○ Subcontractor assessment
	    ○ Performance monitoring and surveillance
	    ○ Environmental monitoring
	    ○ Support incident investigation
	• Describe the involvement of Contractor management in preparing and implementing the following systems:
	    ○ Contract-specific HSE Plan
	    ○ Emergency Plans
	    ○ Training
	    ○ Emergency Exercise/Drills
	    ○ Personal Protective Equipment

Learning and Competence

Learning and Competence

The contractor's learning and competence shall:

- Provide a process for on-the-job orientation for supervisory staff and follow-up of all new employees at the site.
- Provide a description of the procedures (e.g., staff training sessions/ briefings, communication channels, etc.) to take to ensure all affected site personnel, including subcontractor, are aware of the environmental requirements under relevant environmental legislation and permit(s), and contract specifications.
- Provide a method statement on how to provide key site personnel and staff (supervisor and above) with copies of relevant HSE documents, which include but not limited to the environmental management plan.
- Provide a system to ensure new employee training is adequate and coaching in place to improve their working practices.
- Provide a statement on the current standard of workforce and training requirements to meet the contract requirements.
- Provide a training matrix showing all courses (including refresher training), participants (including defined groups and target density), and schedule for the contract, including but not limited to:
  - HSE Orientation
  - Job Procedures (Safe Work Permit, Scaffolding, etc.)
  - Incident Investigation and Reporting
  - Environmental Protection/Awareness
  - Supervisory Training
  - Specialized Training (as applicable)
  - On-the-Job Training
  - Emergency Planning
  - Permitting/Permit Training Requirements
- Provide details of training facilities and demonstrate the availability of suitably qualified trainers.
- Provide a matrix showing records of employees who have been through the program, qualifications gained, and the extent of coverage of training courses.
- Provide a statement indicating worker training before starting work, and that training covers those joining during implementation of the contract.
- Provide a plan for the level of support to provide competent HSE personnel throughout the contract and to indicate in a statement the criteria to select HSE supervisory staff.
- Provide details of the knowledge and experience of HSE professionals. The knowledge and experience should conform to the requirements of the work being executed.
- Provide evidence (i.e., training documentation) that all personnel are trained in applicable procedures.
- Provide evidence (i.e., copies of certificates or qualifications) for all personnel who are required to be certified or qualified

under regulations governing the work site location (e.g., crane operators, fork truck operators, over-sized construction mobile equipment).

Asset Life Cycle

Asset Life Cycle — Not Applicable.

Operations and Maintenance Controls

Operations and Maintenance Controls — The contractor shall ensure operations and maintenance controls include the following:

- A Fall Protection Plan for all activities to be conducted at heights per regulatory requirements (e.g., 6-foot for Canada) where there is a potential for a fall.
- A list of their own work procedures to address management of major hazards and environmental aspects specific to the nature of the SOW.
- A demonstration that their procedures and work practices meet the owner's requirements. Complete a review of owner area- and/or job-specific standards/work practices/rules, which are provided or available through a web access portal such as: www.Owner.com.
- A list of all equipment on the project, identify by type, capacity, and references to standards (as required). Identify the requirements of test certification. Typical equipment includes:
    - Self-contained breathing apparatus (SCBA)
    - Vehicle safety equipment
    - Mobile equipment, including vehicles
    - PPE
    - Environmental monitors (e.g., dosimeters, gas analyzers, etc.)
- All processes that require the use of PPE and statutory requirements.
- Procedures for recording issues, inspection, replacement, and adequacy of stock for PPE.
- PPE requirements for all personnel, with demonstrations of instruction and training provided and evidence of a procedure for checking its use.
- Scheduled criteria for renewing PPE and for re-certification. The organization and person(s) responsible for payment should be identified.
- A list of critical equipment for HSE inspection. The procedure for performing inspections on tools, including those provided personally by employees, and for removing any unsafe tools, should be included as well. Equipment to include (typically):
    - Vehicles
    - Powered hand tools
    - Excavation equipment
    - Scaffolding
    - Lifting equipment
- Developed procedures for on-site cleanliness of work surroundings, such as access to the workplace and lunchrooms.
- A materials/equipment inspection and certification program.

- A statement of methods to ensure that the compressed gas cylinders are properly secured and stored.
- Provide a schedule of periodic inspection of Contractor's work and storage areas to assure safe conditions and practices.

The contractor should ensure regular monitoring of HSE performance and shall complete and submit records to the owner contract manager on or before the due date in the Contractor HSE Stewardship Calendar.

The owner expects full cooperation in the conduct of inspections by the owner, governmental agencies, and other agencies of competent jurisdiction (e.g., Occupational Health & Safety (OH&S)).

- The contractor is to submit copies of citation notices by such agencies to the owner immediately upon receipt.

Contractor (Subcontractor) Management

Contractor Management

Contractor management shall:

- Provide a method for the subcontractor prequalification process.
- Provide a list of contractor-approved subcontractors.
- Provide a list of work activities that subcontractor will execute.
- Ensure the subcontractor meets the same levels of standard as the contractor.
- Provide the contractor's HSE standards.
- Provide a statement that the contractor (and its subcontractors) will not bring a subcontractor on site that is not qualified under the contractor (subcontractor) selection process.
- Apply a verification system to check the implementation of subcontractor HSE standards.
- Provide a list of subcontractor SOWs.
- Provide a description of the subcontractor pre-mobilization process (kick-off meetings, pre-mobilization meetings, performance monitoring, etc.).
- Specify how the contractor's HSE program will integrate subcontractors, paying particular attention to the responsibilities for, and ownership of, the management of transport activities, and how one subcontractor services multiple contractors.

Data, Document, and Information Management

Data, Document, and Information Management

The contractor shall:

- Demonstrate a document management system that includes:
  - How documents are:
    - Controlled
    - Updated
    - Approved
  - Responsibilities for authorization
- Provide details on the use and implementation of safe work permit to use when working in owner facilities.

- Provide contractor HSE procedures, standards, rules, and regulations.
- Demonstrate that a set of rules tailored to the contract is available to all employees.
- Distribute a copy to all employees before starting work in each scope.
- Discuss methods and verify understanding.
- Provide a structure and frequency for updating and disseminating rules.
- Provide an outline of the contractor's record keeping procedure.
- Provide an outline of the contractor's disciplinary procedures related to rule violations.

Contactor Emergency Management

Contractor Emergency Management

The Contractor Emergency Management process shall:

- Demonstrate suitable lines of communication are in place to handle HSE issues in an emergency.
- Demonstrate the ability to communicate with all workers in the event of an emergency.
- Provide a brief description of the potential environmental incidents and emergencies and identify the subsequent response procedures and plans.
- Refer to the owner's emergency management requirements for potential accidents and emergencies.
- Provide a Fire Prevention Program that includes:
    - Restriction on open flame burning to designated areas.
    - Assignment of fire watches, trained and equipped to prevent or control fires, for all welding and burning operations.
    - Proper identification, storing, handling, and use of flammable and combustible material to prevent accidental ignition.
    - Provision of adequate fire extinguishing equipment appropriate for the operations being performed.
    - Training the contractor's personnel in the maintenance and use of such equipment.
    - Evacuation procedures and fire drills as required by owner.

Contractor Communications and Stakeholder Relations

Contractor Communications and Stakeholder Relations

Contractor Communications and Stakeholder Relations shall include the following:

- A project schedule for HSE meetings includes the following:
    - Owner/contractor HSE meetings
    - Contractor safety meetings
    - Toolbox meetings
- A process for maintaining records of personnel attendance
- Meeting agendas, target attendance, and meeting formats
- A demonstration that:
    - Results of HSE activities are communicated to all

    - The meeting program is consistent with the rest of management structure
    - The meetings are recorded
- A demonstration that the use of appropriate communication techniques to make personnel aware of HSE issues, considering any language/literacy problems. Typical examples include:
    - Personal contact
    - Notice boards
    - Field-level hazard assessments
- A description promoting HSE as an intrinsic part of running a business.
- A description of the means for communications on environmental concerns to the owner.
    - Include ways to collect information, report and respond to environmental enquiries, complaints and client's feedback, and ways to disseminate environmental information.
    - The communication requirements should include means of communication, frequency, purpose, and responsible party.

Contractor Quality Assurance

Contractor Quality Assurance	Refer to the Quality Assurance section of the contract.

Contractor Incident Management

Contractor Incident Management

Contractor Incident Management shall include:

- Reporting procedures for:
    - Incident and near-miss investigation
    - Restricted work
    - Reporting injuries (i.e., all deaths, first aid, medical treatment, lost time by personnel, etc.) and health incidents (i.e., disease, exposures to hazardous substances, etc.)
    - Environmental incidents reports (i.e., spills, releases, contamination, etc.)
    - Safety equipment failures
    - Loss of capital equipment and materials
- Details of incident investigation and method to determine root and contributing causes (Are the incident investigation teams led by the relevant leaders?).
- A procedure for follow-up and dissemination of learning points. Methods should be used for collecting incident statistics.
- Details of a modified work program and methods to ensure compliance with the modified work program.
- An effective follow-up procedure to implement the recommendations of its incident investigations, and where applicable those of the owner incident review meeting recommendations.

Contractor Audits and Assessment (Self-Assessments)

Contractor Audits and Assessments	Contractor Audits and Assessment shall:
	• Describe the procedures and requirements of health and safety, environmental inspection, monitoring, auditing/self-assessments, and corrective actions.
	  ○ A matrix showing annual HSE audit and inspection plan should be included.
	  ○ A demonstration that a process is in place for monitoring the implementation and follow-up of inspection and audit action items should be included.
	• Provide a schedule for involvement of personnel in the audit (contractor-internal, joint owner-contractor, and use of subcontractor personnel).
	• Clearly identify in a schedule or plan the frequency of inspection, responsible party, location of the inspection (e.g., equipment lay down yards), reporting, and follow-up of action plan to the owner, etc.
	• Provide details on the involvement of management in review and discussion of findings with the personnel on the contract at HSE meetings.
	• Provide a plan to manage regulatory inspections (i.e., informing the owner) and provide a copy of any regulatory citation notices within 24 hours.
	• Provide details on using the lessons learned to improve performance and implementation throughout the organization.
	• Provide details of the contractor's equipment inspection and certification program.
	• Properly authorize and document a system for rejecting audit findings properly.

Corrective Actions Management

Corrective Actions	The Contractor's Corrective Actions Management Process shall include • Methods for tracking corrective actions from audits, self-assessments, incidents, safety meetings. • Details of compliance mechanism for monitoring/evaluating compliance with the HSE requirements. • A description of procedures and requirements of corrective actions. The owner requires the immediate correction of any unsafe conditions or unsafe acts.

Management Review

Management Review	The Contractor's Management Review process shall: • Determine frequency of management review of the implementation progress.

APPENDIX 8 HAZARD IDENTIFICATION AND RISK MITIGATION FOR SOW

Risk Mitigation Plan

The intent of this form is to ensure the required deliverables have been completed by the contractor (as applicable), prior to the commencement of hands-on work activities.

- Project manager, construction manager, and HSE representative review the list below and identify the applicable requirements.
- Project manager, construction manager, and HSE representative review the list with the contractor.
- All applicable requirements on this Form shall be completed by the contractor.
- The contractor shall attach documentation that illustrates how these requirements are to be implemented and provide the documentation fulfilling the requirements below to the owner.
- The contractor's information shall be verified/reviewed by an owner construction and an owner HSE representative prior to the commencement of hands-on work activities, and shall be kept on file by owner HSE.

A COPY OF THE OWNER'S RISK MATRIX MAY BE PROVIDED TO THE CONTRACTOR TO SUPPORT THE RISK ASSESSMENT OF TASKS AND WORK ACTIVITIES

REQUIREMENT	Applicable		Requirements Met		Outstanding Items		Comments
	Yes	No	Yes	No	Due Date	Responsible	
1. Provide a copy of your Hazard Assessment and Control Plan for the SOW.	☐	☐	☐	☐			
2. Provide evidence of training records for workers.	☐	☐	☐	☐			
3. List the types of activities that most frequently involve worker injuries. Identify the steps that you are taking to reduce the frequency of these occurrences.	☐	☐	☐	☐			
4. Provide a copy of procedures and/ or practices that are applicable to the work activities most frequently involving worker injuries.	☐	☐	☐	☐			

REQUIREMENT	Applicable		Requirements Met		Outstanding Items		Comments
	Yes	No	Yes	No	Due Date	Responsible	
5. Identify high risk activities conducted by your workers and provide copies of procedures that will be used to control these activities.	☐	☐	☐	☐			
6. Provide safety program documents that will be used by workers on site.	☐	☐	☐	☐			
7. Outline your worker competency assessment & development process.	☐	☐	☐	☐			
8. Confirm your ability to directly supervise non-competent workers. Explain the process.	☐	☐	☐	☐			
9. Outline processes that will be used by management to assure compliance with safety program requirements at owner jobsites.	☐	☐	☐	☐			
10. Provide a copy of the system used to identify site safety gaps with associated tracking and sign off for continuous improvement.	☐	☐	☐	☐			
11. Other	☐	☐	☐	☐			
12. Other	☐	☐	☐	☐			

Contract/Tender Number:	
Contractor:	
Work Scope Description:	
Owner Construction/Project Manager:	
Completed By:	
Date Completed:	
Reviewed By:	
Approved By:	

Use the Contractor Risk Assessment Form below to determine the contractor's scope of work (SOW) HSE risk.

- List the primary tasks the contractor would execute within this SOW.
- List the potential loss exposures associated with each task.
- Estimate the severity, frequency and probability of each potential loss exposure. Refer to Appendix A.
- Calculate the Unmitigated Risk Score.
- List all controls that will be used to mitigate the potential loss exposures. Any controls entered that are identified as engineering, procedure, practices, training, special rules or specialized PPE controls shall be awarded a maximum of 0.5 points per category.
- Calculate the Task Risk Score (Unmitigated Risk Score – Controls = Task Risk Score)
- The highest Task Risk Score determines the contractor SOW HSE risk level.

TASKS	**LOSS EXPOSURES**	**RISK EVALUATION**				**CONTROLS**							**Task Risk Score**
List all tasks or activities normally done or that might be done by this service provider.	Most probable loss that would be incurred because of wrong performance or an incident. Consider: Safety, Health, Damage, Fire, Quality or Production problems. Consider: People, Equipment, Materials, and Environment Interactions.	Severity (0-6)	Frequency (1-3)	Probability (-1-+1)	Unmitigated Risk Score (0-10)	List all controls that will be used to mitigate potential loss exposures.	Engineering (-.5)	Procedures (-.5)	Practices (-.5)	Skill Training (-.5)	Special Rules (-.5)	Specialized PPE (-5)	

APPENDIX 9 SAMPLE PREMOBILIZATION CHECKLIST

Contract/Tender Number: ______________________ **Contractor:** ______________________

Work Scope Description: ____

Owner Department Manager:______________________ **Date:**______________________

The HSE Pre-Mobilization Meeting Form shall be reviewed with the contractors' representatives prior to the commencement of hands-on work activities. The intent is to ensure that the contractor is fully aware of owner HSE requirements prior to mobilization, and to ensure the required deliverables have been completed by the contractor (as applicable).

DISCUSSION ITEM	Applicable		Requirements Met		Outstanding Items		Comments
	Yes	No	Yes	No	Due Date	Responsible	
1. Introductions – purpose of meeting/sustainability moment.	☐	☐	☐	☐			
2. Is the contractor an owner HSE-approved contractor?	☐	☐	☐	☐			
3. Is a signed contract for the work scope in place?	☐	☐	☐	☐			
4. Has the 'prime contractor' for this project been established?	☐	☐	☐	☐			
5. Is the contractor aware of statutory requirements and obligations?	☐	☐	☐	☐			
6. Does the contractor understand that they will not bring subcontractors on site that are not qualified under the company's contractor (subcontractor) selection process?	☐	☐	☐	☐			
7. Does the contractor understand the work scope, work location and any related hazards associated with the work location?	☐	☐	☐	☐			
8. Does the contractor understand owner HSE requirements related to the scope of work (SOW)?	☐	☐	☐	☐			
9. Has the contractor reviewed area- and job-specific standards/ work practices/procedures/rules provided to them?	☐	☐	☐	☐			

DISCUSSION ITEM	Applicable		Requirements Met		Outstanding Items		Comments
	Yes	No	Yes	No	Due Date	Responsible	
10. Does the contractor understand the reporting structure and who they report to?	☐	☐	☐	☐			
11. Has the contractor provided the owner with a job-specific organization chart (including names, phone numbers, e-mail addresses of key contacts and resumes of key personnel)?	☐	☐	☐	☐			
12. Have a Contractor Project Specific Hazard Assessment and Control Plan been developed by the contractor for the work scope?	☐	☐	☐	☐			
13. Will the Work Scope-Specific Hazards Assessment and Control Plan be reviewed with the employees performing the work?	☐	☐	☐	☐			
14. Have the Contractor Project Specific Hazard Assessment and Control Plan been reviewed by the owner HSE team?	☐	☐	☐	☐			
15. Is the contractor aware of required environmental due diligence requirements?	☐	☐	☐	☐			
16. Has the contractor identified all hazardous material that will be brought on site (and have they received required industrial hygiene/environmental affairs approval for these products)?	☐	☐	☐	☐			
17. Has the contractor reviewed and agreed to follow the requirements of the Owner Incident Management Procedure (for all injuries, asset damage incidents, environmental incidents/spills, etc.)?	☐	☐	☐	☐			
18. Will the contractor ensure that all employees are English literate and that no one under the age of 18 is allowed to work on the job site? Where workers are deemed non-English literate and are to be used in the execution of the work, the contractor shall prepare a plan.	☐	☐	☐	☐			

DISCUSSION ITEM	Applicable		Requirements Met		Outstanding Items		Comments
	Yes	No	Yes	No	Due Date	Responsible	
19. Will the contractor adhere to the Owner Alcohol and Drug Policy?	☐	☐	☐	☐			
20. Will the contractor adhere to the Owner Smoking Policy?	☐	☐	☐	☐			
21. Will the contractor ensure that all workers adhere to the owner's fall protection requirements?	☐	☐	☐	☐			
22. Is the contractor aware of owner PPE requirements (including welder hardhats)?	☐	☐	☐	☐			
23. Is the contractor aware of owner permitting/permit training requirements/ authorization?	☐	☐	☐	☐			
24. Will the contractor ensure that the following activities take place: pre-job, toolbox, field level risk assessment (FLRA)?	☐	☐	☐	☐			
25. Will the contractor ensure that a FLRA is conducted prior to the start of each task and/or when conditions change or the workers have been away from the work area?	☐	☐	☐	☐			
26. Will the contractor ensure that their Modified Work/ Medical Accommodations Program requirements are adhered to?	☐	☐	☐	☐			
27. Will the contractor conduct a weekly HSE inspection and will they provide the report and follow-up action plan to the owner?	☐	☐	☐	☐			
28. Is the contractor aware of the owner's HSE reporting requirements? The contractor shall:							
28.1. Track HSE statistics, including all of its subcontractor data.							

DISCUSSION ITEM	Applicable		Requirements Met		Outstanding Items		Comments
	Yes	No	Yes	No	Due Date	Responsible	
28.2. Report reactive (lagging) metrics at time of occurrence to the designated owner representative, as per the incident management requirements.							
28.3. Prepare and issue weekly electronic updates to the designated owner HSE representative at a predetermined schedule (typically Sundays before noon), which shall include primarily proactive (leading) measures and total worker exposure hours. Total worker exposure hours include only those worked on owner sites, including all subcontractors.	☐	☐	☐	☐			
29. Does the contractor agree to participate in the owner HSE alignment forum before mobilization to site?	☐	☐	☐	☐			

DELIVERABLES **Contract HSE Requirements**	Applicable		Requirements Met		Outstanding Items		Comments
	Yes	No	Yes	No	Due Date	Responsible	
1. Scope of Work (SOW) Specific HSE Management Plan	☐	☐	☐	☐			
1.1. Project level HSE organization chart **1.2.** Resumes of the key HSE personnel	☐	☐	☐	☐			
1.3. Detailed project-specific roles and responsibilities for each contractor level of worker, including accountabilities and qualifications: management, supervision and worker Detailed project-specific roles and responsibilities for each contractor HSE process	☐	☐	☐	☐			

DELIVERABLES Contract HSE Requirements	Applicable		Requirements Met		Outstanding Items		Comments
	Yes	No	Yes	No	Due Date	Responsible	
1.4. Interface roles and responsibilities between the contractor and subcontractors, site teams, and owner	☐	☐	☐	☐			
1.5. Contractor Management Process	☐	☐	☐	☐			
1.6. Supervision Process	☐	☐	☐	☐			
1.7. Preliminary, high level, hazard assessment and control HSE plan for the scope of work (SOW). **1.7.1.** Identification of the controls, and PPE appropriate to the hazards identified. **1.7.2.** Hazards and risk identified with each work activity, and proposed measures to control risk or eliminate hazard.	☐	☐	☐	☐			
1.8. HSE procedures and practices to be used for this SOW	☐	☐	☐	☐			
1.9. Process for assessing workers literacy in English	☐	☐	☐	☐			
1.10. Temporary Foreign Workers Plan	☐	☐	☐	☐			
1.11. Performance Recognition Program	☐	☐	☐	☐			
1.12. Fitness for Duty Process	☐	☐	☐	☐			
1.13. Drug and Alcohol Process	☐	☐	☐	☐			
1.14. Right to Refuse Unsafe Work Process	☐	☐	☐	☐			
1.15. Program for removing workers from site after HSE rule violations	☐	☐	☐	☐			
1.16. Progressive Discipline Program	☐	☐	☐	☐			
1.17. Process for Permit to Work Authorization	☐	☐	☐	☐			

DELIVERABLES **Contract HSE Requirements**	**Applicable**		**Requirements Met**		**Outstanding Items**		**Comments**
	Yes	**No**	**Yes**	**No**	**Due Date**	**Responsible**	
1.18. Hazardous Product Management Plan, inventory of all WHMIS controlled products, SDS for all WHMIS controlled products	☐	☐	☐	☐			
1.19. Code of practice for Respiratory Protective Equipment	☐	☐	☐	☐			
1.20. Noise and Hearing Protection Management Program	☐	☐	☐	☐			
1.21. Fall Protection Plans	☐	☐	☐	☐			
1.22. Vehicular Traffic Control Plan (for Right of Way & Roadside Work)	☐	☐	☐	☐			
1.23. Contractor Lock-out, Tag-out (LOTO) Process	☐	☐	☐	☐			
1.24. Injury and Illness Prevention Programs	☐	☐	☐	☐			
1.24.1. A manual handling and lifting (body mechanics) program.	☐	☐	☐	☐			
1.24.2. Repetitive Motion Injuries prevention program	☐	☐	☐	☐			
1.24.3. Slips, trips and falls management program	☐	☐	☐	☐			
1.24.4. Heat/cold stress management program	☐	☐	☐	☐			
1.24.5. Exposure Hand-arm vibrations (Hand Tools) program	☐	☐	☐	☐			
1.24.6. Line of fire and hand protection program	☐	☐	☐	☐			
1.25. Risk Management Plan	☐	☐	☐	☐			

DELIVERABLES Contract HSE Requirements	Applicable		Requirements Met		Outstanding Items		Comments
	Yes	No	Yes	No	Due Date	Responsible	
1.25.1. Process used to identify, assess and control or eliminate the identified hazards and associated risks identified in contractor work scope activities	☐	☐	☐	☐			
1.25.2. Action management logs for listing and prioritizing HSE risks with mitigation actions tracked to completion.	☐	☐	☐	☐			
1.25.3. Job Safety Analysis (JSA) or Job Hazard Analysis (JHA) Process	☐	☐	☐	☐			
1.25.4. Field Level Hazard Assessment (FLHA) Process	☐	☐	☐	☐			
1.26. Capturing & Sharing Lessons Learned Process	☐	☐	☐	☐			
1.27. Reliability, Integrity and Calibration of Equipment Procedures	☐	☐	☐	☐			
1.28. Training Process: training and competency assurance Site Access/Training records Training Management Plan and Training Matrix	☐	☐	☐	☐			
1.28.1. Training and Competency Program for Supervision	☐	☐	☐	☐			
1.28.2. Training and Qualification Assurance Program	☐	☐	☐	☐			
1.28.3. Site Specific Orientation for the project	☐	☐	☐	☐			

DELIVERABLES Contract HSE Requirements	Applicable		Requirements Met		Outstanding Items		Comments
	Yes	No	Yes	No	Due Date	Responsible	
1.28.4. Short Service Employees Plan	☐	☐	☐	☐			
1.29. Incident Management Process Incident management and corrective actions processes	☐	☐	☐	☐			
1.29.1. Incident reporting/ notification process **1.29.2.** Investigation Process **1.29.3.** A process to analyze incidents and incident investigation data. Incident tracking system used to track closure of follow-up actions **1.29.4.** A system to communicate lessons learned from investigations. **1.29.5.** A process to ensure HSE alerts are disseminated to all personnel Post-incident notification process	☐	☐	☐	☐			
1.30. Management of Change Process	☐	☐	☐	☐			
1.31. End of shift state of safe work performance safety issues report	☐	☐	☐	☐			
1.32. Internal/External Communications Procedures	☐	☐	☐	☐			
1.33. Site inspections process Inspection Program	☐	☐	☐	☐			
1.34. Audits and Evaluations Program Performance monitoring and reporting process Reviews/audits process	☐	☐	☐	☐			

DELIVERABLES **Contract HSE Requirements**	**Applicable**		**Requirements Met**		**Outstanding Items**		**Comments**
	Yes	**No**	**Yes**	**No**	**Due Date**	**Responsible**	
1.35. Proactive HSE Monitoring Programs	☐	☐	☐	☐			
1.36. Key Performance Indicators – HSE Statistic Reporting	☐	☐	☐	☐			
1.37. Reactive (lagging) metrics	☐	☐	☐	☐			
1.38. Weekly HSE Statistic Reporting	☐	☐	☐	☐			
1.39. Monthly, YTD and PTD HSE Statistical Reports	☐	☐	☐	☐			
1.40. Site visit/tour observation records – Management	☐	☐	☐	☐			
1.41. Site visit/tour observation records	☐	☐	☐	☐			
1.42. Best Practices and Continuous Improvement Program	☐	☐	☐	☐			
1.43. Final HSE Report	☐	☐	☐	☐			
2. Emergency Response Plan	☐	☐	☐	☐			
3. Contractor Security Plan	☐	☐	☐	☐			
4. Environmental Stewardship Process/Environmental Management Plan	☐	☐	☐	☐			
4.1. Specific Area Environmental Management Plans (as required)	☐	☐	☐	☐			
4.2. Heritage, Paleontological and Historical Discovery Practices	☐	☐	☐	☐			

**** If a document section is identified below as applicable, then by default all deliverables of the deliverables listed are applicable, unless noted as N/A.***

Environmental Management Plan Requirements		DELIVERABLES	Applicable		Requirements Met		Outstanding Items		Comments
Document Section	Section or Sub-sections App. or N/A? *		Yes	No	Yes	No	Due Date	Responsible	
1. General Requirements 2. Environmental Management Plan	App.	Environmental stewardship process/Environmental Management Plan (EMP)	☐	☐	☐	☐			
		Conduct where applicable environmental analyses, studies and assessments for all relevant phases of the work: • Additional workspace requirements • Clearing outside the approved footprint • Hydrostatic testing.	☐	☐	☐	☐			
		Other activities warranting regulatory approval such as Pre-Disturbance Assessments (PDA), Temporary Field Authorizations, etc.	☐	☐	☐	☐			
		Environmental Protection and Monitoring Plan	☐	☐	☐	☐			
3. Licenses and Permits 4. Notifications 5. Contraventions 6. Access Restrictions 7. Public Safety 8. Constraint Areas	App.	-	☐	☐	☐	☐			
9. Clearing		Clearing Plan	☐	☐	☐	☐			
10. Grading		-	☐	☐	☐	☐			
11. Soil Salvage		Soil Salvage Plan	☐	☐	☐	☐			
12. Borrow Pits		Practices for borrow pit management	☐	☐	☐	☐			
13. Control of Weeds		Practices for control of weeds/ Weeds Management Plan	☐	☐	☐	☐			
14. Silt Control		Practices for erosion and silt control	☐	☐	☐	☐			
15. Noise		Practices for preventing noise	☐	☐	☐	☐			
16. Water Use		Practices for water use/ Wastewater Disposal Plan	☐	☐	☐	☐			
17. Dust Control		Practices for dust control	☐	☐	☐	☐			

18. Housekeeping		Practices for housekeeping/ Housekeeping Plan	☐	☐	☐	☐			
19. Spill Prevention, Containment and Recovery		Practices for spill prevention, containment and recovery/ Emergency Response Plan/Spill Contingency Plan	☐	☐	☐	☐			
20. Equipment Refueling and Servicing		Equipment Re-fueling and Servicing Plan	☐	☐	☐	☐			
21. Vehicle Water Crossings (Culverts)		Practices to minimize vehicle water crossings environmental impacts/Water Crossings Plan	☐	☐	☐	☐			
22. Roads Water Crossing		Practices for roads water crossing to manage environmental impacts Water Crossings Plan	☐	☐	☐	☐			
23. Pipeline Water Crossings		Practices for pipeline water crossings to manage environmental impacts.	☐	☐	☐	☐			
24. Construction		Environmental protection practices for construction Soil Contingency Plan	☐	☐	☐	☐			
25. Revegetation		Practices for revegetation management.	☐	☐	☐	☐			
26. Site Reconstruction		Practices for site reconstruction.	☐	☐	☐	☐			
27. Clean-up and Reclamation (C&R)		Practices for clean-up and reclamation management Conservation and Reclamation Plan	☐	☐	☐	☐			
28. Waste Management		Waste Management Plan	☐	☐	☐	☐			
29. Contaminated Soils		Practices for dealing with contaminated soil.	☐	☐	☐	☐			
30. Heritage, Paleontological and Historical Discovery		Practices for Heritage, Paleontological and Historical Discovery/Heritage Resources Contingency Plan	☐	☐	☐	☐			

Stakeholder Sign-Off

By signing this form, all parties agree:

- HSE topics listed above have been discussed.
- The contractor is aware of the HSE hazards associated with the scope of work.
- The contractor is aware of the owner HSE requirements associated with the scope of work.
- To comply with all owner HSE requirements.
- Subcontractors used for this SOW shall complete a copy of this form (or its equivalent) to ensure owner HSE requirements are fulfilled.

Function	Print Name	Signature	Date
Owner Contract Owner			
Owner Procurement Representative			
Owner Contract Coordinator			
Contractor Manager			
Contractor HSE Representative			
Contractor Site Manager			

APPENDIX 10 THIRTY-DAY CONTRACTOR HSE IMPLEMENTATION ASSESSMENT

Company:		**Assessors:**		
Assessment Date:				
Scope of Work:		**Basis: Contractor HSSE Plan Implementation Assessment**		
Criteria		**Status**		**Comments**
		Not Satisfactory	**Good**	
Leadership, Commitment and Accountability				
	HSSE Roles and Responsibilities for all positions have been communicated to all on-site employees?			
	All new workers at site have gone through an Employee- on-boarding program. (Green Hands Program).			
	Front Line Management (Site Supervision, Foremen, etc.) implement their own schedule for participation in HSSE activities. (validate participation in but not limited to Toolbox Talks, Weekly Safety Meetings, Inspections and FLHA assessments)			
	Senior Leadership is visible in the field guiding and supporting front line leaders and workers on HSSE requirements			
	Are HSSE issues communicated to all employees (consider Notice boards, Safety Alerts, Toolbox Talks and Safety meetings etc.)			
	Workers know who to ask for HSSE information and guidance			
Safe Operations				
	All materials/equipment have been inspected, certified and stored accordingly.			
	Operations and maintenance procedures are current and accessible to all stakeholders			
	Safe work permits/Toolbox talks and FLRA/ FLHA quality reviews conducted			
Risk Assessment and Management				
	Contract HSE Plan contain the Job Hazard Analysis of scope of work?			
	Field Level Hazard Assessments reviews are conducted?			
	Are the controls in the job hazard analysis being followed?			

	Are site personnel aware of the Environmental Management Plan? Is it signed appropriately?			
	MSDS's for all WHMIS-regulated materials/ products available, communicated and accessible to workers?			
Emergency, Business Continuity and Security Management				
	A copy of the emergency preparedness plan specific to scope of work is readily available			
	At least one Emergency Drill is planned			
Reliability & Integrity				
	Equipment inspection routine in place			
	PMs are taking place on equipment as per schedule			
Personnel Competency and Training				
	Assurance that the frontline management is competent in HSSE			
	Competency and training records are available for all workers on site			
Incident Management				
	Unsafe acts and conditions are reported and documented			
	Contractor personnel demonstrate knowledge of reporting procedures, and Lessons Learned process.			
	Process for evaluating and tracking of corrective actions from audits, investigation and inspections.			
Environmental Stewardship				
	Environmental advisors are available, accessible and visible.			
	Environmental reporting requirements are properly communicated to site leaders, and they are aware of how incidents are managed.			
Management of Change				
	Records of HSSE inspections, audits and assessments are available.			
	Contractor participation in HSSE self-assessment/Audit as per schedule			
Information, Documentation and Effective Communication				
	Contractor Personnel can locate and utilize Safe Work Practices, HSSE Plan, and Environment Control Plan.			

Compliance Assurance & Regulatory Advocacy				
	Records of HSSE inspections, audits and assessments are available.			
	Contractor participation in HSSE self-assessment/Audit as per schedule			
	All pre-mobilization Risk Level 1 (High Risk) deficiencies are addressed and closed or mitigated to acceptable levels			
Design, Construction, Commissioning, Operating & Decommissioning				
	Procedures are available for all critical tasks			
	Construction equipment and machinery are properly sited and maintained			
Contracted Services and Materials				
	Confirm that subcontractor(s) are mobilized by the main contractor according to HFC's requirement. (Prequalification, Induction etc.).			
Performance Assessment and Continuous Improvements				
	Contractor is collecting and stewarding data related to agreed KPIs			
	Contractor submits HSSE statistics as per agreed frequency			
	Performance data shared on notice boards			
	Senior management involvement in review of the Contract HSSE Plan.			

Overall Comments/Expected Findings Closure Due Date:		
Classification	**Level of Concern**	**Management Involvement**
Not Satisfactory	Overall cause for concern	Processes lacked essential control elements or sufficient attention from the business unit. Controlling weaknesses, errors, or compliance issues, either individually or collectively, significantly impair important controls to the extent that prompt corrective action is necessary to correct deficiencies.
Good	Specific	Controls in place are sound and effective based on the risk acceptance and performance goals of management, or any prudent person. Control processes manage compliance with company and third-party expectations. Few, if any, minor errors or deficiencies are noted.

APPENDIX 11 SIX STEPS TO AN EFFECTIVE FRONTLINE VISIT

STEP 1: GETTING READY

- Set a designated time aside to conduct the HSE observation.
- You may wish to arrange to have someone accompany you, as they may aid in the discussion and observation process. They can also give you feedback on how you performed.
- Be aware of the work activities being undertaken at the site of your visit. Think about some of the risks involved with that particular activity.
- Decide the objectives of your visit. E.g., Are you going to:
    - Focus on a particular area or do a general tour?
    - Target specific people for an HSE discussion?
    - Focus on specific work activities?
- Ensure that the site is aware of your visit and that those you wish to provide feedback to are available to meet with you.
- Are any requirements for journey management needed prior to leaving for the site, and have they been put in place?
- Make sure you have all the appropriate equipment with you for your site visit (e.g., notepad and pen, camera, etc.).
- Ensure that you are equipped to comply with the site safety rules.
- Consider taking some Owner token gifts to hand out to workers demonstrating good practices.

STEP 2: CONSIDERING SAFETY AT A DISTANCE

- Talk with workers and leaders about their concerns.
- When walking about on site do not spy—stay out in the open.
- Look at potential hazards from multiple operations.
- Mentally note how people are working and how they may affect each other's safety.
- Are any hazards visible? Is anyone at risk? Are procedures being followed? Is it an environmental noncompliance?
- Look for good health and safety practices (e.g., people working safely, good waste management being applied, etc.).
- Plan your discussion prior to engaging.
- Before you move closer, consider how you could get hurt.
- What change of behavior takes place as you approach? For instance, do workers change the way they are doing the job, quickly put on their PPE, change tools, or stop working? All of these actions may indicate knowledge of the unsafe act.
- Look out for no change in behavior. This may mean workers are unaware of the risk(s) or don't care.
- Formulate your discussion in your head.
- Is it safe to start the meeting?

STEP 3: THE DISCUSSION

INTRODUCTION

- Introduce yourself and explain what you are doing and why. Discuss Owner's commitment to safe and reliable operations.
- Ask for workers' first names.
- Show respect, smile, and be friendly.
- Maintain eye contact as much as possible.
- Ask if it is a suitable time to talk.

WORK THROUGH EACH POINT

- Ask about the job they are doing. How long have they been doing that kind of work?
- What are the risks of that activity? How can they get hurt? How may they hurt others?
- What would their injuries be? What would that mean to them or their families?
- What can they do to reduce that?
- Gain commitment to correct unsafe acts.
- If safe work identified have a similar conversation but commend good safety practices.

SUMMARY

- If possible, get the person to summarize their actions.
- You summarize your actions to support improvements.
- Thank them for their time and involvement.
- Arrange time to call back to review actions.

STEP 4: EVALUATION

THE OBSERVATION

- Stay close to the work site.
- Did you miss any hazards?
- If you did, should you deal with them now or call back later?
- Are you satisfied with the actions?
  - The person at worksite has understood your comments.
  - You have addressed the unsafe act/condition appropriately.
  - Sufficient support exists from the workforce's line management and company. Is the unsafe act/condition a result of a failing in the company procedure as opposed to the worker?
- How firm was the agreement with the worker?
- When should you call back to review and follow up? This is determined by how seriously the person could be injured and how much you can trust them to act.

YOURSELVES

- How did you perform?
- If the visit is performed by a team do not forget to give constructive criticism of technique.
- What improvements to your style and questioning technique can you make?
- Coach each other for continual improvement.

STEP 5: MAKE RECORD OF AUDIT

- Document everything, as omissions may be made if working from memory.
- Keep it brief and to the point (the more you write, the less people read).
- No names of people from work site should be used (unless to commend).
- Record names of auditors.
- Record unsafe acts.
- Record agreed upon actions.
- Ensure the report is sent to the appropriate people.

STEP 6: FOLLOW UP AND REVIEW
• Your responsibility to ensure agreed actions are communicated to the relevant supervisor/ manager for implementation at work site. • Next time you visit, ensure you have a copy of your previous report so that you can revisit behavior is changed. • Commend good safety practices. • Correct unsafe acts (self-correction is the most effective way). • Ask yourself, "Do I need to take any action as a leader?" • Does the organization need to take any action?

Useful Questions to Consider During the Visit

The Job
• Have you been given the site induction course? • Have you been given specialized training associated with your job (e.g., working at heights, lifting & rigging, etc.)? • How's the job going? • How often do you do this work? • How high do you go? • How much ____ do you use? (e.g., cleaner, solvent, pressure, etc.)
How can you get hurt?
• What are the dangers of doing this job? • What do you think could go wrong? • What sort of near misses have you experienced doing this work? • What concerns you about this job? • How could you hurt others?
What would your injuries be?
• What do you think the worst injury would be? • Where would it hit you? • What damage would that do to you? • What would it do to you? How would that affect your family? • What would it be like to have a back injury? • How long would it take for your back to fully recover? • How would the injury affect your work?
What could you do to reduce the risk?
• What do you suggest should be done to stop that from happening? • What could be done to improve your safety? • How do you check this equipment to ensure it is safe before you use it? • When do you do that? How often? • What would you do if you thought there was a problem with safety? • What could the company do to make things safety for you? • Imagine you could decide to improve safety with no restrictions. What would the improvement be? What stops us from doing that? • What could you do to reduce the severity of the injury?

Hazards to Look Out For

There are many ways in which people may be hurt. Below is a list of potential hazards to look out for when conducting a Leadership Walkaround.

Work Site • Trip hazards • Fall from height • Confined space • Poor lighting • Other people above, below, or adjacent • Blind corners • Waste material • Escape routes blocked	**Energy Source** • Electricity • Gas • Radiation • Dropped objects • Temperature extremes • Noise • Stream • Pressure • Explosives
Mechanical • Crushing • Shearing • Cutting • Entanglement • Impact • Stabbing • Friction • Vibration	**Human Kinetics** • Pushing • Pulling • Lifting • Repetitive movements • Over exertion • Static posture
Product Escape • Leak • Spill • Overflow	**Material/Chemical Exposure** • Inhalation • Absorption • Ingestion
At Risk Behaviors • Fooling around • Not wearing PPE • Taking shortcuts • Not following procedure	

APPENDIX 12 PERFORMANCE EVALUATION CRITERIA

Contractor Performance Evaluation	
Contractor:	
Contract Release Number:	
Job Title:	
Date Evaluation Completed:	
Project Description:	
Location:	
Status of Project:	

RATING (5 = Excellent/1 = Poor)	Select 1–5
CATEGORY	
Adherence to Requirements	
Adherence to contract provisions for new hire orientation.	
Compliance with site safety rules.	
Compliance with industrial hygiene requirements.	
Compliance with site contractor's health and safety guidelines.	
Compliance with site permits.	
Safety data sheet (SDS) requirements.	
Response to safety requests.	
Housekeeping	
Monthly housekeeping inspection results.	
Overall site housekeeping practices.	
Chemical hazard control	
Fire control measures	
Flammable and explosive materials control	
General waste—trash collection	
Injury Performance	
Contractor's timely reporting of incidents.	
Contractor's use of modified work program.	
Contractor's participation in incident investigations.	
Number of incidents observed and reported.	
Contractor Safety Program	
On site safety officer.	
All subcontractors employed by the Contractor have been managed effectively with reference to EHS and Quality (meets Owner's standards, practices and/or procedures).	

Safety training for supervisors.	
Substance abuse policy compliance.	
Daily toolbox meeting and/or participation.	
Updates to written safety programs.	
Tools and Equipment	
Availability of tools and equipment furnished.	
Condition of contractor's tools and equipment by site and O&S and Original Equipment Manufacturer (OEM) standards.	
Maintenance of Equipment/Tools/PM Schedules, etc.	
Contractor's knowledge and proper use of equipment.	
Contractor's equipment certification is readily available.	
Contractor's materials inspection/acceptance program followed.	
Costs Within Expectations	
Contractor's cost program.	
Contractor's reasonableness and/or participation.	
Execution of Work	
Contractor's cooperation in handling change orders.	
Contractor's efforts to eliminate nonproductive time.	
Contractor's planning of work.	
Relationship with other site Contractors.	
Contractor's cooperation with Owner and other Contractor personnel.	
Key Field Personnel	
Competence of main office staff and management.	
Quality of first- and second-line supervision.	
Labor Relations	
Contractor's ability to anticipate problems.	
Contractor's efforts in exercising management rights.	
Contractor's relationship with the building trades.	
Project Management/Supervision	
Contractor's efforts to use efficient crew size.	
Control of subcontractors.	
Quality of Craft Labor	
Contractor's ability to obtain qualified personnel.	
Quality of work.	
Schedule	
Contractor's adherence to schedule requirements.	

Quality Control/Assurance	
Contractor's material inspection/acceptance program followed.	
Contractor meets expectations on quality surveillance.	
Issues and Concerns	
Advice regarding potential problems.	
Efficiency in handling paperwork.	
Readiness to accept responsibility.	
Reliability in dealings.	
Response to site requests for action by the supervisor/management.	
Environmental Performance	
Number of spills and environmental exceedances	
Response and clean-up actions for spills	
Readiness to accept responsibility	
Audit Observations and Corrective Actions	
Conducted self-assessment audits	
Respond proactively to audit findings and recommendations	
Other	
Overall Performance Score	

Any specific issues or concerns with the contractor:	
Lowest rated activity:	
Highest rated activity:	
Would you hire this contractor again to do similar work:	
How can this contractor improve its past performance:	
Evaluation By:	
Date Completed:	

APPENDIX 13 SAMPLE CONTRACTOR SAFETY MANAGEMENT KPIS

Leading Indicators

Performance Indicator	No. of Quality Toolbox Talks and Meetings Performed
Method of Calculation	The number and frequency of toolbox talks and meetings to be defined by onsite leadership and will vary based on risk, exposure hours, and requirement to generate improvement. Toolbox talks and meetings are intended to achieve the following: • Definition of scope of work • Confirm resourcing requirements • Verify contractor competency and capabilities • Review of immediate and known risks • Review mitigation plans for risks identified Reporting can be done by tracking the number of job permits signed, and spot check audits on the content by the business unit.
Control Range/Target	100% compliance.
Period of measure	Bi-weekly.
Data collected by	Contractor collects data.
Report by	Contractor to supply and business units' leadership to verify compliance.

Performance Indicator	Percentage (%) of Time of Frontline Leadership Spent in Field vs. in Office
Method of Calculation	Self-reporting by frontline contractor supervisors will be the primary source of reporting. Audits to verify can be done through Outlook Calendar, and consultation with business unit leadership.
Control Range/Target	Recommend a minimum of 50% of time spent in field supervision and oversight.
Period of measure	Bi-weekly
Data collected by	Contractor collects data
Report by	Contractor to supply and business unit leadership to confirm

Performance Indicator	Percentage (%) of Audit Observations vs. Corrective Actions Resolved
Method of Calculation	The field audits would be conducted by business unit HSE personnel. Where risk exposures are identified, corrective actions will be defined, and a timeframe for implementing the corrective actions will be provided. A percentage of audit findings versus corrective actions implemented on schedule should be maintained. Require business units to track and contractor responsibility for executing corrective actions assigned.
Control Range/Target	• High risk corrective actions—100% • 95% of all corrective actions are implemented on time.
Period of measure	Monthly
Data collected by	Contractor collects data
Report by	Contractor to supply and business unit leadership to confirm

Performance Indicator	Percentage (%) of Employees in Competency Assurance Plan
Method of Calculation	Competency of contractor workforce is to be verified during the pre-mobilization phase. Ongoing audits by organization/business unit and self-reporting by contractors must support competency assurance exceeds threshold 75% competency in workforce. It is important to point out that competency extends beyond training and is reflected in competency assessment and sign-off by a competent assessor.
Control Range/Target	Minimum 75% competency in contracting workforce
Period of measure	Monthly
Data collected by	Contractor collects data
Report by	Contractor to provide data and business unit leadership to verify

Performance Indicator	Number of Near Misses Reported with Mitigation Action Completed
Method of Calculation	Daily reviews of near misses reports should be followed up, and corrective actions should be taken to eliminate the hazard and prevent it from re-occurring. The measure would take the percentage of corrective actions taken as a result of the number of near misses reported. Contractors maintain near misses action log and report to organization.
Control Range/Target	Measure
Period of measure	Weekly
Data collected by	Contractor collects data
Report by	Contractor to provide data and business unit leadership to verify

Performance Indicator	Percentage (%) of Quality Job Safety Analysis (JSA) and Field Level Risk Analysis (FLRA) Performed Prior to Beginning Work
Method of Calculation	Number and quality of the JSA and FLRAs should be audited periodically to confirm the analyses are done prior to beginning the work, and these are effective and contribute to reducing the possibility of injury. The analysis should have specific activities and expectations defined so they can be reviewed for quality.
Control Range/Target	100% compliance. Checking the box does not represent quality exercise
Period of measure	Weekly
Data collected by	Contractor collects data
Report by	Contractor to provide data and business unit leadership to verify

Performance Indicator	Balanced Experience in Work Crew
Method of Calculation	This measurement assesses the experience of an assigned work crew. The purpose of this indicator is to ensure that contractor work crews are not entirely comprised of new workers to the worksite or to the work being performed. This leading indicator may comprise two parts: • Amount of time working for the organization at that worksite • The collective experience of the workers in performing the specific work. In both cases, it is undesirable to have an entire crew with limited site knowledge or work experience with the task or work assignment.

Control Range/Target	Average years of experience on work crew. Actual number TBD
Period of measure	During pre-job assessment and crew changeover
Data collected by	Contractor collects data
Report by	Contractor to provide data and business unit leadership to verify

Performance Indicator	**Number of Joint Field Visits Performed by Senior Organization and Contractor Leadership**
Method of Calculation	This is a measure of joint senior leadership visibility. Visits may include a review of the work being undertaken in the field or maybe a specific site safety audit. In all cases of leadership visibility, worker engagement is recommended.
Control Range/Target	Target defined by each contract; 100% compliance
Period of measure	Bi-monthly
Data collected by	Either business unit or contractor to collect data
Report by	Business unit leadership to verify

Performance Indicator	**Number of Safety Alerts Developed (Reviewed) and Shared Across the Organization**
Method of Calculation	This is a proactive measure of the organization's ability to share learning across the business as well as measures the incident reporting effectiveness across both the contractor and organization's businesses. When incidents requiring sharing occur, knowledge generated from investigations should be shared in a way that corrective actions or modified procedures/practices can be put in place to prevent the same incident from occurring elsewhere. Alerts help to facilitate this goal.
Control Range/Target	All high-potential incidents/serious injuries reported 100%
Period of measure	Monthly
Data collected by	Both the business unit and contractor to collect data
Report by	Contractor and organization

Performance Indicator	**Number of Listening Moments and Proactive Safety Improvement Suggestions Submitted by the Contractor**
Method of Calculation	This measure contributes toward the safety culture development of the organization and the level of trust and openness being developed between both contractors and organizations. The measure best reflects the ratio of suggestions received versus action on suggestions implemented.
Control Range/Target	Discretionary
Period of measure	Monthly
Data collected by	Both owner and contractor collect data
Report by	Contractor and organization

Performance Indicator	Percentage (%) of New Contractor Employees Properly On-Boarded and Oriented
Method of Calculation	Proper on-boarding and orientation ensure all workers are aware of the workplace hazards and emergency response plans. Stewardship of this indicator reflects the % of contractor workers who received a site safety orientation as required prior to beginning work on the site.
Control Range/Target	100%
Period of measure	Monthly
Data collected by	Contractor to collect data
Report by	Contractor to provide data, business unit leadership to verify

KPI name	Workmen Compensation Board (WCB) Ratio
Method of Calculation	Worker Compensation Generated
Target	1.0 and less
Period of measure	Per Regulator
Report by	Environment, Health and Safety (HSE)

Lagging Indicators

Performance Indicator	Contractor Total Injury Frequency (TIF)
Method of Calculation	(No. of contractor injuries × 200,000) ÷ Contractor Exposure Hours No. of injuries to include First Aids, Medical Treatment, Restricted Work, Loss Time Injuries, and Fatalities
Control Range/Target	Trending towards Zero – To be determined by the organization based on risk ranking of the contractor category
Frequency of reporting	Monthly
Data collected by	Business units collect data and perform calculations
Report by	Business unit or HSE. Where multiple business units are available, Environment, Health and Safety (HSE) will generally collate to present the organizational performance

Performance Indicator	Contractor Recordable Injury Frequency (CRIF)
Method of Calculation	(No. of contractor injuries × 200,000) ÷ Contractor Exposure Hours No. of injuries to include Medical Treatment, Restricted Work, Loss Time Injuries, and Fatalities
Control Range/Target	Trending towards Zero – To be determined by the organization based on risk ranking of the contractor category
Period of measure	Monthly
Data collected by	Business units collect data and perform calculations
Report by	Where multiple business units are available, Environment, Health and Safety (HSE) will generally collate to present the organizational performance.

Performance Indicator	Contractor Lost Time Incident (LTI)
Method of Calculation	(No. of contractor injuries x 200,000) ÷ Contractor Exposure Hours No. of injuries to include Loss Time Injuries and Fatalities
Control Range/Target	Trending toward Zero – To be determined by the organization based on risk ranking of the contractor category
Period of measure	Monthly
Data collected by	Business units collect data and perform calculations
Report by	Where multiple business units are available, Environment, Health and Safety (HSE) will generally collate to present the organizational performance.

Performance Indicator	Workmen Compensation (Insurance) Ratio
Method of Calculation	Generated by insurance organization
Control Range/Target	<1.0
Period of measure	Per insurance frequency but a minimum quarterly
Data collected by	Contractor receives data and shares with business unit leaders
Report by	Business unit leaders

APPENDIX 14 SAMPLE TEMPLATE—SHARED LEARNING (INCIDENT LEARNING AND PREVENTION)

<table>
<tr><td colspan="4">Incident Learning and Prevention</td></tr>
<tr><td>Site:
Date:</td><td>Technical/Process Safety</td><td>Audience:</td><td>Supervisors and Site Leaders</td></tr>
<tr><td></td><td></td><td>Expected Response:</td><td>Action Per Recommendations</td></tr>
<tr><td colspan="4">Title:</td></tr>
<tr><td colspan="4">Summary of Events:</td></tr>
<tr><td colspan="2">•</td><td colspan="2">Insert Picture Here (A picture tells 1000 words)</td></tr>
<tr><td colspan="4">Root and Contributing Causes:</td></tr>
<tr><td colspan="4">•</td></tr>
<tr><td colspan="4">Key Learnings:</td></tr>
<tr><td colspan="4">•</td></tr>
<tr><td colspan="4">Recommendations:</td></tr>
<tr><td colspan="4">•</td></tr>
<tr><td colspan="2">Source:</td><td colspan="2">ILP No.:</td></tr>
<tr><td>Standard, Policy or Procedure</td><td>Winterization</td><td>Subject Matter Expert (SME):</td><td></td></tr>
<tr><td>Submitted by:</td><td>Job Title:</td><td>Email:</td><td>Ph:</td></tr>
<tr><td>Approved By:</td><td>Job Title:</td><td>Issue Date</td><td>Sequence: 01</td></tr>
</table>

APPENDIX 15 SAMPLE RISK EVALUATION GUIDE (IN THE ABSENCE OF AN OWNER'S RISK MATRIX)

1. Severity: Derived from the losses being incurred as a result of wrong performance of the task; only consider the most likely loss.

Severity Number	Description
0	• No injury or illness or quality, production, environmental, or other loss of less than $ 1000 can occur.
2	• Minor injury or illness without lost time; non-disruptive property damage; quality, production, environmental, or other loss of $1000 to $10000.
4	• A lost time injury or illness without permanent disability; disruptive property damage; quality, production, environmental, or other loss of more than $10 000 but not exceeding $50000.
6	• Permanent disability; loss of life or body part; extensive loss of structure, equipment, material, quality, production, environmental, or other losses exceeding $50000.

2. Frequency or exposure

Number of People Performing Tasks	Number of Times a Task Is Performed		
	Less Than Daily	Few Times per Day	Many Times per Day
Few	1	1	2
Moderate	1	2	3
Many	2	3	3

3. Probability of Loss

Probability	Description
-1	• Low probability of loss
-	• Moderate probability of loss
+1	• High probability of loss

4. Controls: Any controls entered that are identified as engineering, procedure, practices, training, special rules, or specialized PPE controls shall be awarded a maximum of 0.5 points per category.

INDEX

Note: Page numbers followed by f and t indicate figures and tables, respectively

For Product Safety Concerns and Information please contact our EU representative GPSR@taylorandfrancis.com Taylor & Francis Verlag GmbH, Kaufingerstraße 24, 80331 München, Germany